CAREER ADVANCEMENT AND SURVIVAL FOR ENGINEERS

CAREER ADVANCEMENT AND SURVIVAL FOR ENGINEERS

JOHN A. HOSCHETTE

A Wiley-Interscience Publication
JOHN WILEY & SONS, INC.
New York / Chichester / Brisbane / Toronto / Singapore

Library of Congress Cataloging in Publication Data:

Hoschette, John A., 1952–
   Career advancement and survival for engineers / John
   Hoschette.
        p.   cm.
   Includes bibliographical references and index.
   ISBN 0-471-01718-3. — ISBN 0-471-01727-2 (pbk.)
   1. Engineers—Vocational guidance.   2. Career development.
   I. Title.
   TA157.H618   1994
   620'.0023—dc20   93-41343
                                                          CIP

Printed in the United States of America

10 9 8 7 6 5 4 3 2

CONTENTS

PREFACE

During my six years at the university, I received more than sufficient training in science and engineering to be technically successful at my job. Colleges and universities do an excellent job instructing students in the sciences. However in doing so, there is often little or no time left to spend educating students on how corporations function, how people are promoted, and how to survive in the business world of engineering.

Consequently, I naively entered the work force with many misconceptions about how business is conducted. For example, I believed one only needed to do a good engineering job to get promoted. I also believed your supervisor would immediately reward you upon successfully completing an assignment. Another misconceived notion was that someone in the company was always looking out for my welfare. For instance, I thought my supervisor would look out for me just as university advisors did, and my success was the company's primary concern, just as it had been at the university.

Much to my surprise, I quickly learned that business did not operate in this manner. In addition, many of my other preconceived notions were wrong. To get ahead I had to change my thinking and educate myself about the business side of engineering. I had to learn what it really took to advance and then make it happen.

The intent of this book is to help the reader better understand what it takes to get ahead and help him or her understand how companies do business, how they promote people, and provide insights into the dynamics involved in getting promotions.

This book shares some proven techniques that have helped people get ahead and stay ahead. Chapter 1 discusses the importance of getting control of your career. The first step to getting control of your career is understanding the company structure. Understanding the company structure is essential. It can save months on your next promotion, get you out of a dead-end job, and allow you to earn more money. Chapters 2 through 5 review typical company structures.

Next, you must determine the criteria by which you will be judged. Chapter 6 discusses these formal and informal criteria. Formal criteria consist of forms that your supervisor fills out during a progress review. Knowing and understanding these forms is essential for career advancement. However, decisions (influenced by informal criteria) regarding promotions often are made long before the forms are filled out.

Informal criteria are usually never discussed, but are probably as important, if not more important, than formal criteria. Each day your supervisor is evaluating you on the basis of these informal criteria, but do you know what these criteria are? You must find out in order to get promoted. Helpful hints for discovering your supervisor's informal criteria are presented in Chapter 6.

Chapter 7 discusses the use of mentors. What makes a good mentor? What makes a bad mentor? How do you find one? In Chapters 8 through 11, I discuss the things you need to know to manage your career and keep it on track.

Successfully surviving corporate takeovers, mergers and work force reductions is addressed in Chapter 13. Chapter 14 discusses getting on the fast track for advancement and provides helpful hints.

The material here is addressed to graduating engineers as well as those engineers who have been in the work force for several years. My motivation for writing the book stems from two sources. The first is my desire to share what I have learned and successfully applied because no one shared with me many of the insights provided in this book. It would have saved me time and effort if I hadn't learned so many of them the hard way.

The second reason stems from the very positive responses I have received from people who have attended my courses on this subject. This book is an expansion of the material originally presented in a career development course for engineers. The course was developed with the aid of the Human Resources Department at Honeywell over a five-year period to help personnel with their careers. I developed the content with assistance from several company groups and psychologists. The course was taught to all new employees as well as senior people. The material was especially applicable for those who could not understand why they were not advancing in their careers.

The course was also presented to accountants, secretaries, buyers, and marketing personnel. These nonengineering participants found the material very useful, since the principles and guidelines are general enough to be adopted and applied to other occupations as well.

A word of caution about the techniques discussed in this book. The book is not a get-rich scheme. By following its guidelines and hard work, you should significantly reduce the time between promotions and raises. This will mean taking on extra work. Your supervisor will give you more work than you can possibly handle in 40 hours. If you want to get ahead, working over and above your allotted hours is a must.

Your supervisor is not your enemy. He or she has been there before and knows what is going on. Make him or her your ally. Make your supervisor look good and make yourself look good in the eyes of upper management. The techniques suggested in this book only work if you combine them with hard work, good job performance, and honesty. I can only share with you some proven techniques, the rest is up to you.

The ideas and situations which are discussed in the book have been generalized and must be adapted to your company or supervisor. If you try something and it does not work, be patient. It takes time to achieve your goals. If you become frustrated and feel that you should inform your supervisor how things should be done, stop! You have missed the entire point of the book. Your supervisor and/or company has a set way of promoting people, and no attempted revisions on your part will change that. It is your challenge to find out what the criteria are and use them to your advantage.

JOHN A. HOSCHETTE

Lexington, Massachusetts

ACKNOWLEDGMENTS

First I would like to thank my wife of 20 years, Linda, for help in editing the book, putting up with all the inconveniences, missed dinners, and days away on travel over the years. Without her support, encouragement, understanding, and help, this book would not have been possible. I thank my teenage children, Tina and John, for their support by putting up with Dad and all the missed nights teaching classes and preparing this book.

Next I thank all the senior engineers and mentors who have provided me career guidance throughout the years and, in doing so, some of the material for this book. Specifically, John Miller, Don Stevenson, Frank Ferrin and George Hedges, who provided guidance in the early years of my career. Dr. Edwin Thiede, Dr. James Marier and Charles Seashore for their helpful guidance. And Don Willis for his assistance in making a successful career move. In addition the assistance provided by the Honeywell personnel department in sponsoring my first classes and helping with the course material was invaluable.

My parents, Vern and Veronica, for loving support and encouragement throughout the years, I thank them. The Evans Scholars foundation for the opportunity to go to engineering and graduate school. And finally, I thank Joan Manoli for her help in preparing the original draft manuscript.

J. A. H.

CAREER ADVANCEMENT AND SURVIVAL FOR ENGINEERS

CHAPTER 1

ARE YOU IN CONTROL
OF YOUR CAREER?

Are you in control of your career? If you ask most people this question they usually respond with "yes," when in fact they are not. They proceed along day after day assuming that the big raise and promotion are just around the corner. Then when nothing happens they don't understand why. "I deserved it," they say. "Why does everyone else get promoted but not me?" "Why did he get the promotion and I'm still waiting?" The answer is simple: they were not in control of their career as much as they thought and left their promotion up to chance. Never leave your promotion up to chance.

Often you think you are in control but, in fact, the situation is exactly the opposite. You are not in control. Things seem fine according to you and you think the job you are doing is going to get you the big raise and promotion. However, if you look closer and identify all that is lacking, you will soon realize that you have far less control over your career than you thought you had, and far less chance of getting the promotion.

Controlling your career takes a great deal of planning, hard work, and the ability to manage circumstances as best you can to benefit you the most.[1] Knowing everything that it takes to get promoted in your company and constantly working toward that goal is essential to advancement.

There are often many hidden things going on behind the scenes of which you must be aware.

As shown in Figure 1-1, there are several factors that could be limiting your career growth. Key among these are obsolescence, your personality, education level, your supervisor, and maybe even the company or department structure just to name a few.

Always knowing what is happening and controlling circumstances to the best of your ability is key to reducing the time frame between raises and promotions. Are you in control as you should be? Chances are that you are not! To illustrate the point I have developed a few questions which, when answered, usually show that you are not in control as much as you think you are.

1. Was your last pay increase smaller than expected or your performance rating lower than expected?
2. Do you feel your hard work often goes unrewarded? Does it seem other people are always getting awards or recognition?
3. Have you ever been passed over for a promotion and really don't understand why?

FIGURE 1-1 What is stopping your career growth?

4. Do you feel like you are stuck in a dead-end, thankless job with little hope for advancement?

Do you wait until your job review with your supervisor to discuss or plan your career?

If you answered yes to these questions, it points to the fact that you definitely are not in control. You may think you are, but clearly this is not the case. And a yes to the final question only emphasizes the fact that you are not in control of your career as you should be. If you wait until your job review with your supervisor to discuss your career plans, it is too late.

Career planning and preparation must take place long before you enter the supervisor's office.[1] You must be aware of all the dynamics that occur in the company and control them to your benefit that time. Trying to convince the supervisor that you deserve a promotion should be the last thing on your agenda, not the first.

Do you need to be in control at all times? Yes! Yes! Yes! If you are not in control and planning your next career advancement that means you are out of control. Worse yet, you are letting others control your advancement. Failing to take control and plan is simply planning to fail. This is not what you want for successful career development! In addition to taking control, you must also know how to play the game. You must know what to do, when to do it, and how to do it.

DO YOU KNOW HOW TO PLAY THE GAME?

Your fellow employees will tell you simple little things you can do to get promoted: flatter the boss; its who you know; work overtime and be a hero. The list goes on and on. In reality, most people have not done their homework and can only guess. In order to be promoted you must know how the promotion game is played and how to score points. The following example illustrates this point.

Let's assume you are the key player of the team. Everyone is counting on you. The situation is this, it is the second period of the third quarter, the teams have lined up and the goalie is calling the play. On the previous play the putt was good for 3 points, but the right wing was penalized when 5th base was stolen. They have the option to bowl for a strike or fast pitch for a slam dunk. What do you recommend?

Everything sounds somewhat familiar, right? But you really don't know the rules or how points are scored. Therefore your chances of making the right decisions are very small.

If you don't know the rules of the game or how to keep score, can you expect to play well? No!

Getting your next career advancement is similar. You are the key player for this game. The game sounds very familiar—do a good job, brown up to the supervisor, work hard. These all sound like great things to do, but will they lead to career advancement? More than likely not.

If you don't know how your company promotes people or how they keep score, can you expect to get ahead? No!

The bottom line is this: if you don't understand the game you can't expect to score points or do well. Similarly, if you do not know how your company plays the career advancement game, you cannot expect to get ahead. Learn the game. Learn how your company keeps score and start calling the plays that will score you career advancement.

This book will provide you with step-by-step guidelines on how to play the game and score points. It will identify the obstacles limiting your career advancement and show you how to overcome them. It will show you how you can control and determine your advancement. In doing so it will shorten the time to your next promotion or raise. Remember, your next promotion is at least a year away from the time you start planning. So let's move on to getting control of your career and, hopefully, advancing rapidly.

CHAPTER 2

THE COMPANY STRUCTURE—IS IT WORKING FOR OR AGAINST YOU?

The first step in getting control of your career is to determine your company's structures. Usually, there are several structures in your company of which you need to be aware. The first of these is the engineering ladder structure. This structure defines the hierarchy of the engineering levels in the company.

Just as important is the task and reporting structure. Generally there are two types of task and reporting structures utilized by companies. One type is a product-oriented structure and the other is a functional matrix oriented structure [2,3].

Knowing the ladder structures that your company utilizes is essential to career development. Hidden career barriers often exist within these structures of which you must be aware. Learning these barriers and how to overcome them are critical to career advancement. Working within these ladder structures can either be very beneficial or very detrimental to your career.

In this chapter the various company ladder structures are discussed along with the hidden barriers, benefits, and detriments. The key to successful career development is understanding the differences between these structures, identifying what structure your company utilizes and using this information to benefit your career.

UNDERSTANDING THE ENGINEERING CAREER PATH STRUCTURE

A generalized engineering career path structure typical for most engineering companies is shown in Figure 2-1. At the bottom are the nonsupervisory engineering levels. Most companies have between four and five levels of engineering before one reaches the staff or supervisory level. People at these levels are often referred to as the "worker bees," "grunts," or "gofors."

If you are not in engineering, you need to find out what your company ladder structure is. All organizations and professions have structures, secretaries, buyers, accountants, etc. Sit down with your supervisor or someone senior in the company and map it out.

In the lower levels (E1–E2) of engineering you are expected to be a good team player and learn from the senior level, who are considered the old pros. Your assignments are often one or two days long and usually accomplished by yourself or with the help of another person. As you move up the engineering ranks your responsibility greatly increases. In the middle levels (E3–E4) you start to lead small teams to accomplish specific objectives. The objectives are usually well defined. The teams are two to

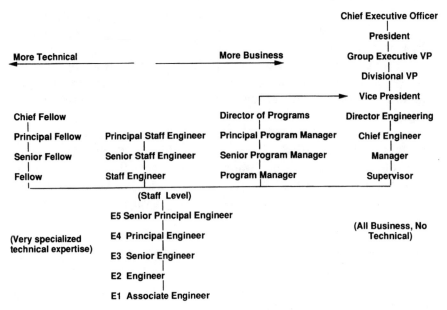

FIGURE 2-1   Engineering career path structure.

three people and assignments may extend over several weeks or months. You are primarily responsible for technical work.

In the upper levels (E4–E5) of engineering you are responsible for directing large teams of engineers with various backgrounds. Often the objectives are not well defined and it is up to you to plan things out. Along with giving the team technical direction you are also responsible for maintaining schedule and cost control. You must have good interpersonal skills to ensure that work is being accomplished through other people rather than doing it yourself.

Once you have risen to the staff level, several career choices have to be made. The career ladder typically splits into four different paths. These paths are shown in Figure 2-1. At the far left is the very technically oriented ladder. This is the fellows ladder. These people are usually PhD's with a very specialized expertise. They are often the recognized experts of the company. They typically are known outside the company and throughout the industry for their expertise. Stereotypically they are thought of as having silver hair, a messy office with patents, plaques, and awards hung up. Usually they are interested in only one thing—science! Fellows usually only deal with technical issues and avoid management or cost issues whenever possible. Learn who they are in your company. They know thousands of technical shortcuts and can usually get you out of a jam if you don't understand something.

A word of caution about fellows. They have seen practically everything and are very thorough. Don't shoot from the hip around them; they will blow you out of the saddle if you do. Be precise and exact. Associating with fellows can accelerate your career or dead-end it quickly. If they are well respected it helps but if they are labeled as eccentric professors it can be very detrimental.

The next ladder to the right in Figure 2-1 is the staff engineering ladder. These people usually have their master's degree and deal with putting systems together. The staff engineers work with system-type problems such as overall performance of the products, for example, the overall mileage performance of a car. This may include aerodynamic shaping, motor efficiency, weight, and speed, whereas fellow engineers will usually deal only with one specialized characteristic of the product, for example, the performance of the spark plug.

Staff engineers, like fellows, concentrate mainly on technical issues. However, with their broad background they often serve in advisory roles to management on technical as well as schedule and cost issues of the project. They are very analytical persons who are responsible for the big picture on how things are coming together technically. You can usually recognize them by the fact that they typically have the best computers in the group. Their office walls sport collections of flow charts and graphs showing trade studies. They will try to put everything into one huge equation and often talk about budgets. If you try to get an answer out of them, they will always qualify it with the phrase "That depends upon . . . ."

A word of caution about system engineers. System engineers analyze the big picture. They want to know what all the causes and effects are. So if you go into their office with only the small picture they might quickly label you as not understanding the problem. Being labeled incompetent is exactly what you don't want your boss to hear. Be prepared if you plan on interfacing with them.

The next ladder to the right in Figure 2-1, is the program manager ladder. These are people who typically run programs or projects. They are responsible for cost and schedule performance and have little to do with the technical work. They organize teams of engineers to accomplish the tasks on their programs. They generally are the primary interface to the customer. They are easy to recognize: most program managers have huge schedules hung up in their offices. The schedules usually show every task planned and the progress made.

The program managers control the funding for programs and have little to do with personnel and salary administration (raises). The program managers are focused on getting the job done on time at the cost they quoted. They are not interested in personnel problems and often the company will pay them a bonus if the project is brought in on time and within cost.

A word of caution about program managers. They are not interested in excuses. Their bottom line is get the job done. You may find that they can be abrasive at times, principally, I believe, because their main focus is the job (work to be done). They can do a lot to help you get your raise, or prevent it.

Finally, the engineering career ladder to the far right in Figure 2-1 is the management ladder. These people are responsible for profits and personnel. They hand out the raises, promotions, and demotions. They make most of their decisions based on the bottom line (profits). Their philosophy is "if we can make money lets do it; if not, stop doing it!" I've seen great engineering projects come to an end quickly because they were not making money for the company. This is usually the tallest ladder of the company and the highest salaried.

The company engineering ladder is discussed first to let you know that it does exist, to make sure you are aware of it, and to get you thinking about your career path. Second, as your career develops and you move up the ranks, you will have to decide which ladder fits your plans, whether you want to be technically or business oriented. As you move up you will need additional education and training. You can shorten the time it takes to move up the ladder by getting the appropriate training before you advance. If you like technical work, you should prepare yourself by taking more technical training as you grow. If you desire a more management-oriented career, then you should take more business training. There is no right or wrong career path. There is only what is right for you.

As you move up the ladder each level becomes more demanding. You must decide for yourself what level is right for you. If you have a family and family is important to you, you must decide if becoming Vice President is right for you. Usually, to become Vice President the family must suffer. Some people stop at the lower levels, others at the middle levels. For these people this is comfortable and a good choice. How high you decide to climb on the corporate ladder is strictly a personal decision that everyone must make for themselves. But beware—the higher you go the more responsibility you accept.

Another choice that people must make as they ascend the ladder is whether to stay technical or go into management. Staying technical means your career interests lie along the fellow and system-engineering paths. Your primary focus is technical advancements. Deciding to move into management means your career interests lie along the program manager or management paths.

Having a successful career does not only mean becoming president of the company. You can have a very successful career and remain technical. Your success is not measured in terms of titles, power, and profits. It is measured in terms of breakthroughs, papers published, patents, and technical awards.

The bottom line here is that you need to define what success is for you. If it is getting to level E3 and you make it, then this is success for you. If it is becoming a supervisor and you make it, then this is success for you. Success for you may not be measured in terms of titles, raises, or promotions, it may be as simple as doing the best job you can. Not everyone can become vice president of the company, nor do they want to. However, everyone should reach their own definition of success.

Now that you have had time to review and understand all the levels, which do you think is the most difficult one? In my opinion it is the first-line supervisor—your supervisor! Everything in the company comes together at this level. All the great ideas generated by the president of the company on how to run things must be implemented by the first-line supervisor. All the policies geared to running the company must be followed by the first-line supervisor—vacations, health care, drug testing, expense reports, employee training, security, raises and promotions, project successes and failures, personnel problems, and department budgeting to name just a few.

Put simply, all the direction from management above falls on them and all the headaches from employees below rise up to meet them. They are the single point in the company where everything comes together. They must do everything and it all has to be done now. They are often overloaded with work. Make it easy on them instead of harder. They're trying to do the best job they can, so give them a break. Often, most first-level supervisors quit and return to the technical ladder rather than continuing on in the management ladder. A high percentage of managers end up divorced. It is a very stressful and time consuming position.

Make your supervisor your friend, not your enemy. The last thing they need is you running into their office every time something goes wrong. Remember, when you go in to talk to your supervisor, he or she probably has a thousand things on their mind.

Let me make several interesting observations about the engineering ladder structure. Between the lowest level and the top there can be as many as 12 levels. Not all organizations have 12 levels in the corporate structure. Some have only 8 or 10; others may have as many as 12. In any case, lets look at how long it would take you to get to the top if there were 12 levels. If you could get a promotion every 3 years (and this is considered very fast rising), it would take you 36 years. Therefore, you had better make everyone count. In addition, look at the odds. For a major corporation, the president is usually 1 of 40,000 people. This roughly makes your odds 1 in 40,000, which are not very good.

An easier way is to know someone at the top and have him help you rise to the top. This is called sponsorship and, believe me, it happens. The alternative to coping with all this structure would be to start your own company, but that's another story for another time.

Most companies require that their top level executives have a broad background and understand all aspects of the business. For this reason, most executives will have worked on several different career paths before they make it to the top. Most companies require them to spend time doing program-management jobs and a wide variety of other jobs before promoting anyone to a presidential level. In many companies, it is the program-management ladder that leads to the vice presidential level.

Invisible barriers exist all along the ladders. One barrier that stretches across all ladders is the higher-degree barrier. Shown in Figure 2-2 is the typical pyramid structure of the company with the Chief Executive Officer (CEO) at the top and the workers at the bottom. I have taken a general survey of the types of degrees that people have at various positions in the company. Below the first-line supervisor, about 80% of the people have bachelor degrees, about 15% have master degrees, and about 5% have doctoral degrees. In between the first-line supervisor and the director level, about 50% of the people have doctoral degrees, 40% have master degrees, and 10% have bachelor degrees. Between the director level and the CEO, about 60% have doctoral degrees and 40% have masters degrees. Often at this level a director or VP will have multiple degrees: a doctorate in engineering along with a masters degree in business. In my survey I found that there was no one at these upper levels with only a bachelor's degree.

So the message comes out loud and clear here. If you want to become an upper level manager someday, you will need an advanced degree. There are invisible barriers based on educational level in most companies. The

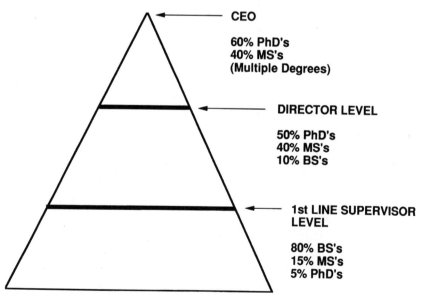

FIGURE 2-2   Hidden educational barriers in a company.

only way you are going to find this out is by asking around about the degrees various people have in your company. This should reveal two interesting things. First, how prevalent advanced degrees are in your organization and, second, the type of advanced degrees that virtually ensure getting ahead. If you are a mechanical engineer working for an electronics company and all upper level managers are electrical engineers, that tells you about the promotion policies being followed. Similarly, if you are a software programmer in a chemical company and all the advanced degrees are chemical, it tells you something about how far you can go.

In any case, investigate to determine what types of degrees people have. There is nothing more ego inflating, i.e., flattering, than asking the vice president about their background while at the company party. They will immediately tell you everything they have accomplished. This opens your door to opportunity. Ask what they would recommend for a younger employee. Hopefully, at that point you are hearing the shortcuts that you need to know and their hidden criteria for promoting.

GETTING AHEAD IN PRODUCT-ORIENTED ORGANIZATIONS

In addition to the engineering ladder structure, other ladder or organizational structures exist that you must be aware of. These organizational structures define the roles and responsibilities of the people involved as well as the manner in which engineering tasks are accomplished. Most companies utilize teams of engineers organized into either product-oriented structures or functional-matrix structures. A typical product-oriented organization is shown in Figure 2-3.

In the product-oriented organization everyone works on the same product. Often the entire department is responsible for getting the product out the door. Everyone in the product-oriented organization reports to a single upper level manager at the top of the organization. The product-oriented organization in Figure 2-3 illustrates the reporting organization for the ZX50 car. Reporting to the departmental manager are the individual subsystem departments, in this case, four different departments. For the ZX50 car there are the engine department responsible for designing and testing engines, the body-design department responsible for the exterior body, and so on. Each of these departments performs a necessary job within the organization.

Reporting to the department manager is often the program manager and

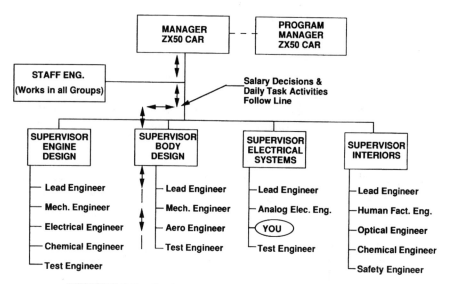

FIGURE 2-3   Typical product-oriented organization chart.

usually one or two staff engineers. The program manager determines what is to be worked on and the engineering department determines the best technical approach. The work to be accomplished is planned by the department manager and the program manager. One nice thing about this type of organization is that your work direction and salary review often come through your supervisor. One person, your supervisor, provides the work direction and hands out the raises.

Although in agreement as to the outcome, the program manager and engineering department manager may not see eye to eye. The program manager wants the quickest, fastest job done and the department manager wants to do the best engineering job. Often this means doing opposite things on a project, but always with the same goal.

There are several advantages to working in a product-oriented organization. First, the organization gets to build the entire product. It is very rewarding to see the entire product come together in your department. It gives you a real sense of accomplishment. When the product is a huge success, management knows whom to reward. Second, everyone reports through the same chain and decisions are more easily made in this type of organization.

Third, you often work with people who have very different backgrounds, and gaining new perspectives from them greatly broadens your background. Since staffing of organizations tends to be fairly lean, you might be the only one in the group with your type of background—an excellent and advantageous position, since there is no competition. Another advantage is that if you have the same background as the supervisor or manager, they can more easily appreciate and understand the work you are doing. Managers tend to promote people with backgrounds similar to their own. The logic for this is: It worked for me and I'm successful, so why not continue the trend?

There are some disadvantages to working in a product-oriented organization. When the product development stops, so does your career. Also, if the product runs into trouble (due to poor planning or performance), chances are you are going to be labeled as coming from a failed program.

Another disadvantage of the product-oriented organization is the fact that your boss may not have the same technical background as you. If you are a chemical engineer and he is electrical, it will be difficult for him to appreciate the great job you are doing. And, finally, when the product comes to the end of its development, there is no place for you to go unless a new product is being developed. Working on a product often creates engineers who know everything about one little thing, a difficult position to be in if you plan on running the entire organization some day.

GETTING AHEAD IN THE FUNCTIONAL-MATRIX ORGANIZATION

A typical functional-matrix organization is shown in Figure 2-4. In this structure, the company is organized around functional groups. Everyone in the functional group usually has the same background but may not work on the same product. The left-hand column of the functional-matrix structure identifies the departments in the company according to function. The top row of the matrix identifies the projects within the company. For the example shown, three types of vehicles are produced: the ZX50 and ZX90 cars and the S5 truck. The top row of the matrix identifies the program manager(s) responsible for getting work done.

In this type of organization the supervisor usually determines the salary increases and the program manager usually determines the work to be accomplished. Companies utilize the matrix organization since it is a very efficient way to run the business of large organizations. Highly skilled people of a particular background are grouped together. When a program manager needs an engineer with a particular expertise he can usually have

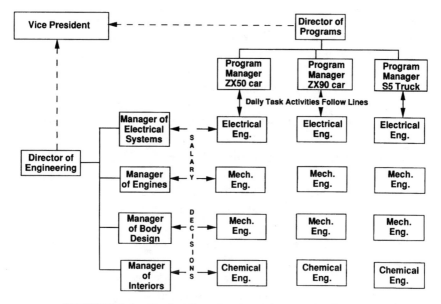

FIGURE 2-4   Typical functional-matrix organization chart.

one assigned to his program. This type of structure allows the company to quickly move people on and off programs.

Working in a functional-matrix organization offers several advantages for your career. First, you are working with people who have a background similar to yours. This is advantageous when you get stuck on a problem. Usually someone in your department has faced a similar problem and is readily available to help you. Second, you may work on more than one product at a time. This can really help if one of the products is very dull work. Third, in a functional-matrix organization you can quickly move from one product to another as they come and go.

Working in the functional-matrix organization also offers some disadvantages. Most of the time work direction comes from your program manager and technical direction comes from your supervisor. This places you in a very tricky situation when the program manager and your supervisor disagree. If you follow your supervisor's direction, it may make the program manager upset and he could very possibly kick you off the product. If you follow your program manager's direction, it may upset your supervisor and hurt you at raise time, since he directly controls your salary.

In a functional-matrix organization you really have two bosses to please on any given day—your supervisor and the program manager. You have to find a way to keep both happy if you are going to get that raise. Often the program manager and department supervisor sit down and discuss your performance to determine the raise you will be getting. The ultimate decision is up to your department supervisor, but you had better make sure the program manager agrees with his assessment of your worth.

It doesn't hurt to simply ask each of them how you're doing. It's better to find out before rather than after raises have been given out, while you are still able to do something about it. Another disadvantage to the functional-matrix organization is that you are placed with a group of people who all have similar backgrounds. Its tough to shine or stand out in the crowd when everyone there is doing the same thing. You have to look for ways to be different.

Most companies are not strictly organized in either a product-oriented or functional-matrix fashion but rather are a mix of both. In order to move ahead in either organization you must know how it works and what the pitfalls are. No one organization type is better for your career than the other. The important thing is to recognize there are differences and that what works in one may not work in the other. Find out how your company is organized and who determines your raises. You could be working hard to impress someone in the organization who has no involvement with your raise. Make sure you report your progress on a continuing basis to the person who controls your raises. In this way he will know what a great job you are doing.

Recent graduates often come into a company thinking that someone is looking out for them. They go into a corner, do a great job, and when they are finished expect that their boss will reward them. Remember, no one is looking out for you but you. Don't leave your next promotion or raise up to chance. Spend time each week with your supervisor, let him know the problems you are solving and get feedback on your performance. It may be uncomfortable to hear some criticism as to how you are doing, but its the only way you are going to turn things around.

In summary, we have discussed the engineering, technical, and business ladders often used in companies. In addition, the hidden education barriers of these ladders have been highlighted. The two types of the organizations—product oriented and functional matrix—have been reviewed along with the advantages and disadvantages of each.

The next step in getting control of your career is to study and know the engineering process that goes on in your company, which ultimately gets product out the door. In Chapter 3 we will discuss a generic engineering process that usually occurs in most companies and identify how knowing this process can benefit your career.

ASSIGNMENTS

1. The first homework assignment you must complete is to map out the engineering ladder in your company. A good place to start is with your boss. Inquire about the levels within the company, next ask to see the organization chart and where you fit in.

2. If you can, obtain a description for each of the levels. Usually, every company at one time or another has written down the job description for each level. It is interesting to compare the descriptions as you move up the ladder. Often the descriptions for the executives are not available. But judging from the job descriptions from several levels below, the executive job descriptions must include walking on water, talking to God, and working several miracles a day.

3. Obtain a copy of the job descriptions for your level and the next level up. First study the description for your current level and see how you compare to it. Remember, if you want to get promoted you'd better make sure you are fulfilling every requirement. Next study the level above yours and see how the levels compare. We will discuss this later in the book.

4. Determine how your company is organized. Is it a product-oriented organization, a functional-matrix organization, or a combination of both. Who reports to whom. Can you identify all the levels and the person at each level between you and the vice president? If you can you're in great shape! If you don't know who they are, how can you expect them to know who you are and promote you. In later chapters I will discuss how to get visibility at the upper levels.

5. The final homework assignment is to determine what your career path will be. How high up the ladder do you want to go and how much time will it take? What actions must you take to reach your goal?

CHAPTER 3

THE ENGINEERING PROCESS IN YOUR COMPANY—LEARNING THE ROPES

The engineering process is simply the steps that companies go through to design, build, test, and deliver products for their customers. Understanding this engineering process for your company is an essential prerequisite to career advancement. Once you have learned the process used by your company, you can use this information in a multitude of ways to benefit your career.

For example, by knowing the engineering process you can quickly identify the departments that are critical to the company's success. Are you working for a critical department that is absolutely necessary to the company's survival? Or are you working for a department that is only there to solve a short-term problem and will disappear shortly after the problem is solved? People in critical departments get raises and promotions. People in short-term departments get reassigned or laid off.

Another benefit to knowing the engineering process in your company is that you can contact other departments well in advance and schedule their help for your product. This allows you to get things done more efficiently, for less money, and on time—results that motivate companies to give raises and promotions. By not learning the engineering process you are doomed to trial and error methods that usually cause delays and cost overruns—results that do *not* lead to raises and promotions. Remember, no one likes to have an unexpected job dropped on them without notice or time to respond. By going to these departments well in advance, you give them the opportunity to schedule and complete your job in a timely manner. This makes both the company and you look good.

Knowing the engineering process will aid you in determining what projects are best to work on and what projects should be avoided. If you are assigned to a project that has nothing to do with the mainline processes of the company, you can be assured that you will also probably not be in the main line for career advancement.

By studying the engineering process and learning all the steps involved you quickly become aware of the possible shortcuts and ways around all the red tape. In doing so, you should be able to accomplish assignments in a much shorter time. Getting assignments completed sooner and for less cost than expected usually results in career advancement.

A side benefit to learning the engineering process is that you will also educate yourself about the company's products. Knowing both the company's products and the engineering processes provides a valuable insight into the company, a valuable insight which you can use to your advantage. For example, once you learn the products of your company, you can determine which product is best to work on for the betterment of your career.

For a good example illustrating this point, let's assume you are working on the highest-profit product in the company and you just figured out a way to produce it 10% cheaper. Chances are very good that someone is going to notice your improvement and reward you for it. Or you may be working on the most critical calibration and assembly process for the most profitable product of your company. Having knowledge of this is always great leverage when it comes time to ask for a raise. Supervisors like to reward key personnel who are absolutely essential for getting products out the door.

The point here is that you will not find out about these positions unless you study the engineering process and products of your company. You must analyze the process to determine what are the critical steps and what products are the most profitable. The optimum situation for your career is being responsible for the key process steps of the most profitable product of your company.

Failure to study and learn the engineering process and key products of your company will only sidetrack your chances for career advancement. A good example of this is working on a product that is about to become obsolete and phased out. Under these circumstances you can pour your heart and soul into the job, but the chances for promotion are not very positive. In fact you may be shortly facing a layoff.

Another example of how ignoring the engineering process can hurt you is when changes are introduced into the process. Often companies will change the engineering processes to eliminate unnecessary steps, or they may send out the work to other companies that can do it cheaper. If you are working on a process step slated for being phased out, this will only hurt your chances for career advancement. Only by thoroughly understanding the process will you be able to determine how changes will affect you. With this knowledge you should be able to sidestep trouble and keep your career on track.

Knowledge of the engineering process is essential for successful career advancement. It provides you with so much valuable insight into which jobs are critical and which jobs are not, which process steps are important which steps are not.

The engineering process is different from company to company. Not only is it different from company to company, it is continually changing within any one company. Therefore you must be continually updating yourself on the changes. Never assume that once you have determined the engineering process for your company you are finished. Learning and understanding the engineering process in your company is a never ending career activity. Now that you understand the extreme importance of learning the engineering process, let's look at all that is involved in the engineering process.

Generally, it takes a team of people of various backgrounds to get a product out the door. The size of the teams vary anywhere from 15 to 150 people depending upon the product. In addition, it takes several different departments working together to design, build, and test the products. It is extremely important that you identify the different departments and people who become involved in producing your product and the order in which it is produced. You must determine all the steps a product must go through in your company to get out the door. Next diagram the steps. This is often called a product flow diagram. They normally exist in most companies, but you may have to do some research to find one. If product flow diagrams are not available, then you need to generate one.

If you don't know how to generate one, ask! Ask your supervisor, as he generally has a good feel for what it takes to produce the product. Listen closely to what he tells you, for he will be sharing with you how he thinks things should operate. I am certain he will also tell you about all the pitfalls he encountered and how he overcame them. Remember, you have two ears and only one mouth—you should be doing more listening than talking.

After you have mapped out your company's engineering process you should quickly come to the realization that 95% of the work done on your product is actually performed outside your department. The engineer must call on the support of other departments to help him design, build, and test the product. What this means to you is that you must have good inter-personal skills, as the utilization of work done by others is extremely important. This brings us to a very important question.

How well do you interface with other people? Can you easily convince other people to support you? If you find yourself having problems with this, don't panic. You can get help developing interpersonal skills. Guid-ance or career advice is available through local colleges and universities. There are plenty of evening classes offered to show people how to deal with these problems. Contact them to obtain a course listing. Remember, career advancement depends on how well you interface with people. How good are you at requesting and receiving their help?

When dealing with other departments, always try to create a win–win situation between your department and the department you are requesting help from. A win–win situation is one in which you get what you want and the department doing the supporting work gets what it wants. In other words, you both look good in accomplishing the work. Here's an example: you bring your job to the support department early enough so that they have time to respond and accomplish the work on time. You look good because you planned ahead and the work was completed on time and within the budget allocated. They look good because they were able to complete the work on time and with high quality.

Win–win situations usually result in promotions or career advancement.

Stay away from a win–lose situation. This is where you win and the supporting department loses. A typical win–lose situation is where you are late bringing your job to the support group. You may even have to go over the head of the support group to get top priority for your product so it is completed on time. The support group rearranges its priorities and gets your impossible job done, but this throws all other jobs in the department behind schedule. You win because you get your job done, but everything else looks bad and the support group loses. These win–lose situations will always come back to haunt you on the next project. That support group is going to put your job last and possibly keep it there to get even. Believe me, I have seen this happen.

> Win–lose situations usually result in demotions or no career advancement.

Sometimes projects require you to create a win–lose situation between departments. There are several ways around this situation. First get the department supervisor to agree to rearrange the priorities in the department. Try not to go over his or her head. Stay within the department to resolve the conflict. Raising it to higher levels usually creates more problems than it solves.

Something else you can do is submit the person who did the work or the department for a company award. This helps to smooth over ruffled feathers and will make it easier the next time you have to deal with the department. Still another suggestion is to highlight to upper management how the support department is handling your impossible situation. Nothing makes a support department move faster than knowing upper management is watching their efforts. It's their time to look good and they want upper management to see them at their best. Finally, don't take credit for their work. Nothing makes people more irate than someone taking credit for their work. Make sure you give credit where credit is due.

> If a support group did all the work, they should get the credit, not you!

Learning all the products and processes of your company is not an overnight task. It may take you months and maybe even years before you fully understand all the products and processes. Don't become discouraged, stay with it. You will find it was worth the effort.

A short cut to learning the process sooner is to study your company's "Policies and Procedures Manual." All larger companies have one. This manual will define the policies and procedures that are to be followed by the employees regarding normal business operations. It is a simplified guide to how the business does business. It maybe extremely boring reading, but it is well worth the effort.

To summarize, the benefits of understanding your company's engineering process are:

1. By identifying the steps that products go through during design, production, and testing, you gain valuable insights into the company's operations.
2. Understanding enables you to determine what products, functions, and departments are critical to the company's success.
3. Knowledge of other departments' functions leads to better productivity.
4. Awareness that business requires a cooperative effort.
5. Realization that 95% of the work is done outside your department.
6. Realization that interpersonal skills are important—the key to getting people to support you is through win–win situations.

WHAT DOES THE TYPICAL ENGINEERING PROCESS INVOLVE?

A typical engineering process is shown in Figure 3-1. This engineering process has been generalized and will vary from company to company as well as from product to product within any one company [2,3,4]. The example was chosen since it is fairly typical of most companies. This example represents a starting point for determining the engineering process in your company. Study the example, then use it as a guide for your developing your flow diagram. Mark it up and change it as you need to.

The engineering process starts with a product idea. If the product idea is worth pursuing, the company forms a team. The team will write a proposal to build a particular product for a certain customer, to improve a product line, or to develop a new product. The team includes engineers, program managers, proposal writers, accountants, and marketing or sales people. The marketing people provide information obtained from the customer about how the product should be built. This information includes the customer's specifications and requirements, the time frame for completion, and the customer's budget.

With this information the team formulates a proposal to build a product to meet the customer's need for the least amount of money and deliver it within the time constraints. The engineers design the product and describe its operation. The accountants compute all the costs associated with building the product. Getting everyone to agree as to what should go into the proposal is no easy task.

The proposal is reviewed by the customer for its technical merit, cost,

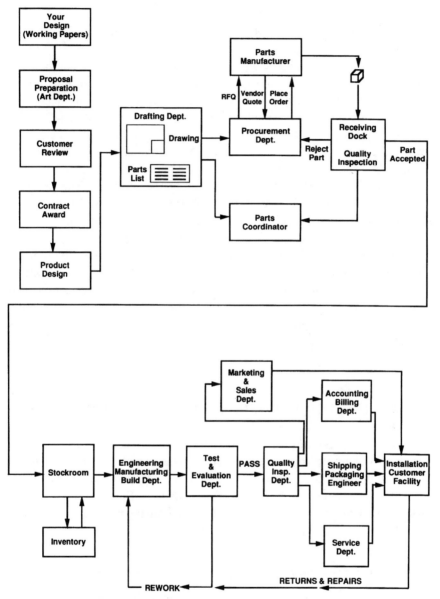

FIGURE 3-1 Block diagram of a typical engineering process.

and schedule. This review process can take anywhere from several weeks for simple contracts to months and even years for large developmental programs.

If the proposal is accepted and your company is awarded a contract, the next step is the product design phase. For the product design phase, management organizes a team of engineers to develop the product. Often the teams are organized around a Program Management Office (PMO). It is the responsibility of the PMO to execute the contract and make sure work is carried out as planned. These PMO's are lead by a director who has profit and loss responsibility for the program. The director will have program managers to assist him in running the program. The program managers are responsible for organizing the work. They determine the tasks to be accomplished, the order in which they are to be done, and the schedule for getting the work done for the design engineers.

After the program has been thoroughly planned out and the team members identified, the design work is ready to start. This step is known as product design. During this step, every detail of the product is designed and performance modeled. Design trade studies are performed showing different ways to build the product. The pros and cons of the various methods are identified. During the design phase the engineer is utilizing everything he learned in school and then some.

GUIDELINES FOR BETTER PRODUCT DEVELOPMENT

The design process varies from company to company and from product to product. I can only share with you some generalized guidelines that will help you design a better product. The following design guidelines are very useful and should apply to most design situations. Simply reviewing this list as you develop your design should help eliminate mistakes and identify problem areas.

Guidelines for a Better Design

1. Before you start the design, write down all the requirements placed on the design or product. Summarize them in a consolidated table or matrix so it is easy to review. This list should include as a minimum:

Performance requirements	All inputs or outputs
Clearly defined interfaces	Size constraints

Weight constraints	Power constraints
Safety constraints	Test requirements
Operating constraints	Customer use
Special conditions	Rework requirements
Reliability (MTBF)	Human engineering
Safety	Maintainability

2. Get agreement on the product requirements from your system engineer and/or supervisor. This is a must! If you design your product to the wrong requirements you will stand little chance of having it approved and you will waste valuable time and company money.

3. Generate a list or table of different approaches that you might use for the design. The table should contain acceptable and poor points for each design. Remember there is always more than one way to do the design. By generating different approaches you can make trade-offs to determine which design is best. Also, by generating different approaches you are not locked into only one.

4. Now show the various design approaches to people in your organization and find out what they think. You will get some good tips along the way. Also, by showing your approaches to people, you will get a feeling for what is good and bad. You should show these approaches to your supervisor. If he likes one and disapproves of others, you have a good indication as to which approach to take. It is best to find out which approach stands the best chance of approval rather than spend a lot of time and energy on an approach that will never be approved.

5. Analyze every detail of your design. Model everything you can about the design. The modeling should quickly identify the problem areas for you. By modeling the expected performance of your product, you can show how the product meets every requirement on your list. Engineers will often start reinventing the wheel by generating all their own models. Before you start modeling, find out what models exist in the company. A good place to start checking is with the senior-level engineers in the company. Chances are they already have a model you can adapt to your needs with minor modifications.

6. Build a mockup of the design if possible. A mockup usually does not function but has the correct form or shape. It will help you to visualize how the product comes together. It has been said that one picture is worth a thousand words. I have found that one mockup is worth a thousand pictures! Mockups can identify problems in advance and help you correct them early in the design process.

7. Identify contingency plans in case something goes wrong. If you base your design on a specific part and the part is suddenly no longer available, what are you going to do? Check to make sure all the raw material and parts are available to build it the way you have designed it. A good practice is to find two different suppliers for each part. This way if one goes out of business or no longer produces the part, you have a backup.

8. Keep track of product costs as you design it. A well designed product is no good if you cannot sell because it is too expensive.

9. Before you commit to the design, develop a build and test plan. This will help you to quickly identify whether you have all the resources to build and test the product. If you don't have the resources, you had better let the company know in advance so they can make arrangements to get them.

10. Put plenty of design margin into your design. If your product must survive a five foot drop, design it to survive a seven foot drop. If your product must operate within 5 seconds of turn, design it to operate within 3 seconds of turn. Design as much margin into your product as it will allow you. This can save a great deal of expensive and unnecessary redesign later on.

11. Write down and document every aspect of your design. Keep good design notes. You are the only one who has a complete understanding of the design but remember your co-workers must build and test it. They need good notes and documentation to do this. It is better to make a mistake on the side of too much documentation than too little. Take the time to make sure your documentation is accurate. Check all numbers twice.

12. Build a full function prototype model before you start to build the final design. These are often referred to as breadboards or brassboards of the product. A prototype will allow you to see any problems in advance and give you time to change things. It will also alert you to possible potential problems.

Remember, these guidelines are very general and you have to adapt them to your particular company or product.

During the design phase, the engineer may have to go through one or more formal design reviews. The intent of the design reviews is for the engineer to present his design for review to make sure he or she is meeting all the requirements and that all parts will fit together when the product is assembled. Design reviews are often attended by senior engineers and management in the company. Their purpose is to review the design and ensure that ideas will work. They will provide helpful hints and suggestions on how to improve your design. They are also making sure that lessons learned from other products are being applied and that mistakes from other products are not repeated. Often the customer will participate in the design review.

These design reviews are often referred to as Preliminary Design Reviews (PDRs) and Critical Design Reviews (CDRs). The PDRs are often held at the beginning of the program to make sure it gets off to a good start. At PDRs the engineers present to the customer the plans for development and a preliminary design for the product. Later on in the program a CDR is held. The purpose of the CDR is to officially mark the end of the design phase and the start of the build and test phase. At the CDR, the final design for the product is presented and approval is received from the customer to start the actual building and testing.

Design reviews are an excellent opportunity to shine. A well prepared PDR will shorten the time to your next promotion. It will make your boss look good and the team look good as well. It's worth the extra effort to make sure everything has been accounted for and the presentation is well organized. This is the time to show them your best. On the other hand, a poor PDR can significantly hurt your chances for a promotion. It's exactly the wrong time to look bad. So keep this in mind when you are preparing for a design review.

At most design reviews there is usually a lot of heated discussion on what the correct approach should be. Your design will receive a lot of helpful criticism whether you like it or not. Remember not to take the criticism personally. Remain calm! It's difficult when it's your hard work being criticized. But remember, others may have more experience and usually have good suggestions. The sign of a good senior designer is not taking the criticism personally and remaining calm. The best thing to do is find out from others how they would improve it and then use their suggestions. This way accomplishes two things. First, you get them to participate in the design (this is especially good if a customer is participating). Second, you have shown that you are cooperative and can deal with change. Both are excellent qualities for engineers to have. If you really want to impress people, thank them for their suggestions! Everyone likes to hear the words "thank you." It makes them feel as if they have contributed something useful, a feeling even the most senior-level executives enjoy. Finally, when you are a senior engineer reviewing the design of a junior engineer remember how it felt when you were in that position. Choose your words and criticisms carefully. You should be honestly trying to help the junior engineer and not just trying to make your importance known to others.

Once the design is complete, the hand sketches and rough drawings are brought to the drafting department to draw up the parts of the product. The drafting department documents the design by creating a drawing package. The drawing package contains a complete set of drawings that document every part in the product, how it is built, and how it is to be assembled. Drawing packages may contain anywhere from 10 to 10,000 prints, depending upon the complexity of the product.

If you have an opportunity to choose with whom you work in drafting, try to get the most senior draftperson. The reason is that he or she has seen many designs in his or her time and knows what has worked and what has not. The senior draftsperson has a multitude of little tips to make the design more producible, perform better, and have a better chance of succeeding. If he finds something that improves the design, highlight this—tell everyone, especially his boss, about his contribution to the design. Believe me, I've seen the drafting department find a lot more of my mistakes once they knew they were going to get credit for it, mistakes missed by the senior engineers and myself.

When completed, the drawing package usually goes through some type of official sign-off before it is archived in the company print room. The transfer to the print room is usually referred to as a drawing release. To release drawings there are several engineering standards utilized in industry. These standards define what should be on the drawing and the levels of control. The control levels define who should review and sign it before it is released as well as who has authority to change it. At the lowest level of control, it is signed by the engineer and draftsperson only. At the highest level of control, the draftperson, the engineer, his supervisor, the program manager, quality control, manufacturing, and even the customer review and sign drawings.

Most engineers look upon a drawing release as a genuine nuisance and avoid it if they can. My recommendation is to not avoid it, but use it to your benefit. Once all your drawings are complete, schedule some time with your boss to go through them and sign them off together. This is an excellent opportunity to show him or her all that you have accomplished. A neat, well organized drawing package is very impressive. It also gives you a chance to show all the problems you've solved, how any suggestions made were incorporated and how you have things under control. These are all things that can highlight your contribution as an employee and shorten the time to your next promotion.

After drawing release, the prints are used by the procurement department or "purchasing" as it is sometimes called. The procurement department uses the drawings to obtain bids from other companies to build the parts. The procurement person contacts companies that are interested in building the part and requests a quote from them. This is referred to as a Request For Quote (RFQ). The manufacturer sends the bid back to the procurement department for review and approval. Usually the procurement department and the engineer decide to whom the bid will be awarded. When a decision is made, a purchase order is placed with the chosen company.

Getting the purchase order signed-off is an exercise that will try any engineer's patience. A typical purchase order will usually require the signature of four to five different people—your supervisor, your supervisor's boss, your program manager, your procurement agent, and the parts manager, just to name a few. On larger programs it may also require the production engineer, the quality engineer, and the program accountant.

For those parts fabricated within the company a drawing is brought to the department that does the fabrication and the department normally starts work immediately. No purchase orders are necessary. This is much simpler, so most engineers will try to fabricate everything in-house.

To keep track of all the parts being fabricated most program managers utilize a "parts coordinator" to keep track of everything. The job of the parts coordinator is to place the orders, track delivery dates, push the parts through incoming receiving and inspection and get them into the stock room as quickly as possible. If you want to have your parts built first it doesn't hurt to make friends with the parts coordinator. Remember, all the other engineers on the program are also trying to get that same parts coordinator to order their parts. Parts coordinators can make or break you. If you get on their good side, they can make sure your parts always get top priority. If you get your parts first, you stand a better chance to finish first. If you get on their bad side, you may never get your parts. I have had coordinators tell me, "See that purchase order there, well the engineer was just yelling at me to get it out. Too bad! It's going to the bottom of the pile. It will get processed in about two weeks. That will teach him."

The point I'm making here is that other people in the company can make or break you. You need to understand the function of each department and how it can affect you. You need to stop and think, if I win the argument and make an enemy, am I really winning?

You can make enemies with one or two departments in the company and survive, but if you make enemies with any more, chances are you will not survive in the company. Too many people will be out gunning for you. There is a saying, choose your friends well. It's also good to choose your enemies well so they can do the least amount of damage to your career. Chances are you can make enemies with some departments and your career will not suffer. But make an enemy in a department that is key to getting your job done and you may be greatly limiting your career.

Parts ordered from outside the company are usually logged in and sent to inspection/receiving. In the incoming inspection and receiving department the parts are usually inspected to make sure that all parts were built to the required specifications. Most companies do this for two reasons. The first is to detect bad parts before they are used to build the product. This can significantly reduce rework later on. Second, it gives the company a much stronger case for return and free replacement of bad parts by the manufacturer. This is especially true if lawsuits can be the result of poor inspections.

In order for the quality inspector to examine the incoming parts they must know what to look for. To get this answer, the inspector usually obtains a copy of the part drawing from the company print room. This is called inspecting the parts to drawings. These are the same drawings that you had the drafting department create and release. As most quality inspectors say, "No print to inspect to, then no parts get through." It pays to have all your prints completed and released prior to inspection.

From receiving inspection the parts usually go to the stockroom where they are logged in and stored. Make sure your parts get to the right place once they leave the inspection department. Parts have been known to get lost between the inspection department and stockroom. Most of the time the parts coordinator will take care of this.

KNOWING WHAT TO DO WHEN IT COMES TIME TO BUILD YOUR PRODUCT

Before starting to build your product, it is good to generate a check list of everything that you might need during the build. Some companies actually hold a build readiness review prior to the start of building the product. The following is a good check list to go through prior to commencing the build.

1. All parts have been received and are in the stockroom.
2. A build procedure has been written showing the build process flow step by step.
3. All inspections and tests to occur during the build are defined and

agreed upon prior to start of build. Data sheets for recording results of inspections and tests are available.

4. All hazardous steps have been identified and people informed of any dangers.
5. Technicians are available and trained for every step of the process. Training or practice on scrap parts for the more difficult assembly steps is a good way to reduce rework.
6. All necessary documentation (drawings) is available for the technician.
7. Persons to contact in case of problems have been identified.

These are some of the questions that should be asked and answered prior to start of build of your product. You will have to generate your own list and tailor it to your company's or product's need.

Most companies have an engineering build or manufacturing department that actually builds the product. These departments are often referred to as "model shops" for small quantity builds. This is where a breadboard or brassboard model of your product is first fabricated. Using the prints and assembly procedures that you have developed, these departments quickly assemble the product. There are two rules of thumb during assembly that can help you get through the build process easily and quickly. First, be available for the technician so she can ask questions on how you want it assembled. Remember, if you are not around, the technician will assemble it the way she thinks you want it. This can lead to a lot of mistakes. Second, be ready to make changes if necessary during the assembly process. Chances are that you did not think of everything while designing it and the first place this will show up is during assembly.

It's best if you are on the spot and can quickly remedy the situation. If you are not around and the technician has to go to your boss or other co-workers to find out how to solve the "glitch," it doesn't look good and can hurt your career. Solve problems quickly and as soon as you can. Spend as much time as possible in the assembly area during this time.

Just as in the drafting department situation, request the assistance of the most senior technician if possible. He has seen a multitude of products and knows hundreds of shortcuts or helpful hints. Make him your friend and not your enemy. Remember, some technicians like to point out all the mistakes of the design without mentioning its good points. It is very tough to sit there and listen to how bad your design is and not take it personally. Remain calm! Draw out helpful suggestions and incorporate them if you can. Be sure to give credit where credit is due. Again, it does help if you point out to the technician's boss all the good ideas that were contributed and how he or she is helping the design.

At the completion of your product build, write up any lessons learned from the build process. Be sure to update your build-process flow with any changes you have made to it. Updating and improving the build-process flow is extremely helpful the next time you plan on building the product. Make sure you share the lessons learned with your supervisor and other people in the organization. This may save them from going through the same problems in the future. Also, as others share their lessons learned, you will learn faster and more efficient ways to build problem-free products.

KNOWING WHAT TO DO WHEN IT COMES TIME TO TEST YOUR PRODUCT

The test and evaluation phase of the program follows the build process. The best way to prepare for this step is to generate a complete checklist of everything that should be done to test the product prior to shipping. The following is a generalized checklist that should help you in preparing your own.

1. Generate a test-requirements document that identifies parameters to be measured, requirements of test equipment, and pass/fail criteria.
2. Generate a test plan that calls for verifying every requirement listed in the design specification you generated. Get inputs from test personnel on the test that you plan to run.
3. The test plan should show the order of the tests to be performed.
4. For each test to be performed have the following been identified?
   Test facility available
   All test equipment in place and calibrated
   Test objectives for each test identified
   Data sheets for recording results completed
   All test, quality, and inspection personnel notified
   Test procedure written, reviewed, and approved
5. Identify the test director or someone responsible for each test.
6. Establish contingency plans should failure occur during the test.

A good engineering practice is to witness and monitor all aspects of testing. The engineer should be readily available to answer questions as they come up and provide direction to the test team as required. In addition he or she should be comparing the test results against the modeling results obtained from the design-modeling phase. A summary of the documentation identified in steps 1 through 3 is shown in Figure 3-2. Generating this documentation is highly recommended as it will be needed in successfully controlling the testing phase.

Generally, there are four types of performance verification methods that are utilized during the test phase. These performance verification methods are:

1. *Analysis.* This is usually a mathematical modeling of the product to show compliance with requirements. (For example: safety analysis of product handles.)

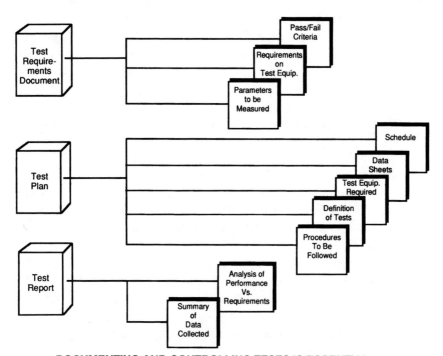

DOCUMENTING AND CONTROLLING TESTS IS ESSENTIAL

FIGURE 3-2   Test documentation.

2. *Inspection.* This is usually a physical and visual inspection to verify performance of the product. (For example: inspection of labels on the product to ensure correctness.)
3. *Certification.* This is verification of performance by receipt of certification from manufacturing. (For example: certification on how pure certain chemicals were that were used in the build.)
4. *Test.* This is verification of product performance by operation and/or measurement of an item, usually requiring instrumentation to record and evaluate measured data. (For example: measuring the weight, size, or power consumption of a product.)

Note: Often it does not seem important to know exactly how the performance of a product is verified. However, if one considers the costs associated with the verification method, one quickly realizes how important the test method can be. Do not forget the cost factor associated with each test method. Usually tests are most expensive and certifications are the least. Make sure you know the costs of performing the different tests in your company and minimize the cost of testing where ever possible.

In addition to defining what types of tests the product will be subjected to, there are different groups or combinations of tests that a product may be subjected to prior to shipping it to the customer. These groups of tests often include:

1. *Design verification or development tests.* These are groups of tests that engineering performs on the product to verify all aspects of the design. Most often this is the largest group of tests. Some of the tests may be performed only once; since all the products are designed exactly the same, all others will pass it. Other tests in this group may be performed several times before the product is shipped and/or repeated for every product. (For example: measure the power consumed by each television set.)
2. *Qualifications tests.* These usually are sets of tests that challenge the environmental performance of the product. (For example: vibration testing, drop testing, humidity testing, temperature testing, etc. . . .)
3. *Acceptance tests.* These are groups of performance tests that the customer usually specifies to be run on every product prior to shipping.
4. *Burn-in tests.* These are groups of tests that the customer usually specifies to ensure that no instant or short-term failures will occur.

These tests may include combining several tests into one. (For example: operate for 50 hours while cycling between hot and cold temperatures.)

A very good engineering practice is to develop a matrix showing exactly how and where every requirement of the specification will be met. An example of this is shown in Figure 3-3. The column on the left identifies the product-specification requirement. The next four columns identify the test method used to verify the specification method. The next four columns identify the tests to be run during the various test groups. And the final column identifies the test-plan paragraph that describes the test to be performed.

This type of table makes an excellent communication tool. It provides a condensed summary of all testing that is going to happen. It is great for the test team to have and can help clearly identify to the customer what is going to happen. Figure 3-3 is a generalized table that you should adapt and tailor to your product or company.

Finally, after the tests are completed, a test report and lessons learned

Specification Paragraph	Test Method				Test Group				Test-plan Paragraph Describing Test
	Analysis	Inspection	Certification	Test	Design	Qual. Test	ATP	Burn-in	
3.1 Power On/Off				X	X	X	X	X	4.1 Power On/Off
3.2 Paint Finish		X				X			4.2 Paint Finish
3.3 Material			X			X			4.3 Material
3.4 Hazard/Safety	X				X				4.4 Hazard/Safety

FIGURE 3-3 Requirements verification matrix.

summary should be written. The test report should contain a summary of the data collected as well as an analysis of performance versus requirements. Lessons learned should identify solutions to problems that were discovered during the test phase of the program.

Taking the time to review test results with your boss can be very beneficial to your career. Take the time to explain to him how the product performed against predicted results and the lessons learned. Make sure you point out to him the problems you overcame and the improvements you identified for future projects. Don't overwhelm him with data. Make sure it is neat and organized. Remember, he is forming ideas about your performance and your abilities. Take time to polish the report; now is the time to show him your best.

After successfully completing the testing of the product it is time to ship it so that the company receives payment. Once you are ready to ship several other support organizations get involved. As Figure 3-1 indicates, marketing, accounting/billing, shipping, and the service or installation departments get involved. Make sure they complete their job in a timely fashion so you can still meet the schedule.

Here are some tips that may help company morale and benefit your career at the same time. First, before you ship the product, get photographs of it for future customers. Second, if you can get the team together, have a team photograph taken. Give copies of the photograph to each team member. See if you can have the company newspaper publish a story about the team and the product. Get a picture of the customer receiving the product if you can. And, finally, submit people who have helped you for awards; they will be more willing to work with you the next time.

The engineering process discussed in this chapter was a very general one. You can use it as a starting point for mapping out the engineering process in your company. By generating a map of the engineering process you will have a better understanding of how your company does business as well as what products and functions are key to its survival. By knowing the steps in the engineering process you can contact other departments well in advance and schedule their help for your product. This allows you to get things done more efficiently, for less money, and on time—all good reasons for raises and promotions.

You should have realized by now that 95% of engineering is done outside your department and you must depend upon others to help you. Do you have the good interpersonal skills needed to make this happen? If not, take some classes. When dealing with other departments try to create win–win situations; stay away from win–lose situations [5]. Pushing a product through large companies or organizations can seem impossible at times. Several senior engineers have shared this advice with me on occasions when I felt like giving up. Their words were "Don't let them wear you down; when you know you are right, keep on going."

The engineering design, build, and testing is a complex process. No one person can do it all but with the help of teammates you can do it. Develop guidelines or checklists for each step to help you ensure that everything will be done. By generating guidelines or check-lists jointly with your teammates they get involved with the work and help you more. Let support groups make contributions to the product to get their commitment; you'll need it. Who knows, their contribution may be just what is needed to make the product really successful.

Don't ignore documentation. This is a very important part of the design process. It helps you keep track of the design as changes are made, it provides a means to communicate to others all that is involved in your design and it provides a record of what was built after the product leaves the company. Overdoing the documentation never hurts; underdoing the documentation always causes problems.

Finally, the engineering process changes continuously. Departments come and go. Policies within departments change, people change, and new ways of doing things are continuously being implemented. So always be watching for how the engineering process changes. It is part of your job.

ASSIGNMENT

1. Map out the engineering process in your company. How does your department fit in? Can you identify a name or person for each department or function in the engineering process?
2. What are the key products of your division? Rank them according to the profits they generate.
3. What are the critical engineering process steps for the product you work on? Who controls these steps?
4. Write up a specification for your product. List all the requirements.
5. List the models you have available in your department which you can access to help you model your product's performance.
6. Generate a requirements summary and test matrix for your product.

CHAPTER 4

GETTING THE MOST FROM YOUR COMPANY'S EDUCATIONAL SYSTEM

From the previous chapters one can see that higher education is an absolute must for career development. However, obtaining further education is not a simple undertaking, nor does it guarantee career advancement. You must make absolutely sure that the large amount of time you will be investing in further education will pay off. This chapter identifies how to use and get the most from your company's educational system. Helpful guidelines are discussed to ensure that your further education will be appreciated by your company and will result in career advancement for you. Education for education's sake does not benefit anyone. Choosing the correct type of further education for you benefits everyone.

WHY FURTHER EDUCATION IS A WIN–WIN SITUATION

Most companies gladly support further education for their employees. There are several reasons for this. First, an employee who has received further education is often exposed to new and improved methods. These new methods may allow him or her to solve problems more quickly and easily. Applying these new methods to the job can result in *cost savings* to the company or better products. The cost savings can be significant and easily pay back the employer's educational costs. Education usually has a good return on investment for the company. Second, educational expenses can be recouped by the company since they are tax deductible.

A third reason companies support further education is shared knowledge. Often times, a company will send one employee to a class. When the employee returns he must share with his fellow employees the new knowledge he has gained. Thus the company can effectively educate many employees through the training of only one person.

Another reason companies support employees returning for further education is the contacts the student may make at the classes. Often classes offered at technical symposiums are attended by people with similar backgrounds looking for solutions to their problems. The other students in the class may even be the company's customers or competitors. By attending these technical symposiums and making contacts, an employee may learn what the competition is doing or may encounter future customers. Thus the company may learn what the competition is doing or may even develop new customers. These are nice side benefits to allowing the employee to take a class.

Finally, most large multimillion dollar engineering contracts require a company to have university professors on staff acting in an advisory role. These consulting professors provide the high level technical expertise often required to successfully complete the project. Therefore, sending an employee to take courses and establish one or more contacts while at the university is extremely important to the company.

For all these reasons companies gladly offer further education or tuition reimbursement plans for their employees. The company benefits and the employee benefits.

> Further education is a win–win situation—you win and the company wins.

Your first challenge once you have decided to return for further education is to find out how your company's tuition reimbursement plan works. The best starting point to answering this question is your supervisor. Contact your supervisor and discuss with him your ideas for returning for further education. Watch for his reaction. If he thinks it's a good idea, you now have one of the most important people in the company supporting you. If he objects to you returning, find out why. Maybe his department has no budget for you. In this case, don't panic; try to get him to budget for your education in the next planning cycle. He may also say no because the timing is bad and he needs you at work. If this is the case, discuss with him when the timing would be better and start your plans there. Make sure he knows that you will not let him down while you are returning for your education.

Your supervisor may object for still another reason—insecurity. She may feel threatened by your returning to school. If this is the case, go slowly. Your supervisor will probably not tell you this but continue to make up a thousand and one excuses why you cannot take the classes. If you sense this is the case proceed cautiously. This is a major roadblock that should not be there. Give your supervisor time to adjust to the idea. Maybe this is what she needs. You might also point out that by giving you the opportunity to pursue further education, she is really developing your career and that is part of the supervisor's job. Developing you as a junior employee she will look good and it should be considered a feather in her cap. If none of this works you will have to make a job change. You cannot return to school and have your supervisor not support you while you are attending school. It simply will not work.

As difficult as it may sound, you may have to change supervisors or departments. If you should change supervisors, search out a supervisor who will support you with encouragement and funding. In any case, just remember:

> Do not take no for an answer. It's your career!

Often your supervisor may not be up on all the latest rules and regulations regarding your company's tuition reimbursement program. Your next contact should be the human resources or personnel department. The personnel department is usually responsible for administering the employee tuition reimbursement program. Meet with personnel as soon as you can and discuss the details of your company's plan. Learn everything you can about the program. Here is a sample list of the things you might inquire about.

1. What employees qualify for the program? Do I qualify?
2. Is there a limit on tuition reimbursement costs? Do they cover tuition, books, etc.?
3. Am I required to pay for classes in advance? What conditions must I meet to qualify for reimbursement?
4. What courses and majors apply for tuition reimbursement?

5. Are there restrictions on the universities or schools I can attend?

6. Am I allowed to miss any work time, and if so, can it be made up?

While you are at the personnel department make sure you get the names of co-workers who are also using or have used the program. Contact these people immediately and meet with them if you can. The company cafeteria during lunch is a very good place to do this. Meet with as many people as you can and discuss their experiences. Any helpful direction or shortcuts that they have learned could save you hours of time later on. By contacting other people who have used the program you may find someone who is returning for the same courses that you plan on taking. Their help could be most valuable in setting up your program. Or they could help you with classes they have already completed. Having the names of other people in the company you could contact or draw upon for support during tough times is a must to survive.

Normally the company tuition reimbursement program requires the completion of several forms. These forms will require information about the courses, expenses, and explanations of how the courses relate to your work. (Courses must be related to your work in order for the company to claim the expense as tax deductible.) Forms can scare people away from ever getting started. Don't let them scare you. In fact, if you handle the situation correctly, filling out the forms and getting approval can actually help advance your career.

The tuition reimbursement forms will usually require several levels of management to sign and approve. The reason for this is to make sure that managers are aware of the tuition costs that will be accruing and be able to budget for them. Don't ask anyone to get your forms approved for you. Do it yourself by scheduling time for a one-on-one meeting with each manager who must sign. This gives you an excellent opportunity to meet your managers. These will be the same people who will be approving your raises and promotions.

One-on-one meetings with the managers who must approve your forms is an excellent way to get visibility. Go to these meetings well prepared and ready to discuss your further education—what courses you are planning to take and the benefits you see coming from the courses. Ask them if they have taken any similar courses. Or you might inquire who the manager recommends you contact when you need help. Not that you expect to need a lot of help, but planning ahead is a good thing.

If you are required to obtain the signature and approval of your supervisor's supervisor, use the time wisely. Be sure to point out all the help that your supervisor has been, how the class will benefit the company, and how you plan on sharing what you will learn with others in the company. Remember, after each person has signed and given their approval, shake their hand and thank them. They too will be investing a significant amount of money in your education, and they at least deserve a "thank you" for it.

DIFFERENT TYPES OF CONTINUING EDUCATION (PROS AND CONS)

Choosing the best type of continuing education for you will depend upon your long range career goals. There are various forms and types of education that you can pursue. Advanced degree programs leading to a master's or PhD are offered through most major universities in the United States. The starting point is contacting your local university and requesting their extension or night school programs. Programs usually exist for advanced degree programs in both engineering and business.

Another avenue available to receive graduate level education is through satellite courses. Universities such as MIT and Harvard use satellites to transmit their courses throughout the country. Check with your company to see if they have a satellite hookup available and receive classes. This is an excellent way to take graduate courses, since you may not even have to leave the facility.

Further noncredited education is available through seminars often offered by experts throughout the country at various times during the year. The best sources for finding out what type of seminars are being offered are the company library and trade journals. These seminars are usually intense week-long seminars taught by experts from the universities or industry. Their purpose is to provide the student with the opportunity to quickly "come up to speed" on the state of the art.

Another noncredited method of obtaining further education is by attending symposiums offered by the various engineering or business societies. By joining an engineering society, you will automatically be sent information about upcoming symposiums and training seminars. These symposiums provide excellent reviews of fundamentals as well as knowledge on the breakthroughs. They offer the opportunity for you to meet with people of a similar background and exchange ideas. You can find out what other people across the nation are doing to solve problems. They also provide a very good means for new job contacts. During these symposiums, many companies post their job openings. You can quickly compare what other people are making and which companies are growing and expanding. This is an excellent means to find out what other career opportunities are available for you.

Finally, some companies offer in-house training or courses given by experts from within the company. This training usually consists of short courses taught at lunchtime or after hours. Contact your supervisor or personnel department to find out if any internal courses are planned. Instructors of in-house courses make excellent contacts to have within the company. There is no better way to meet the company's experts than in a nonthreatening one-on-one after hours course. After taking the course, these in-house experts become valuable resources for you to call upon in times of trouble. They also become another vote of confidence for you. When new engineering teams are being staffed, management often calls upon their experts to make recommendations for staffing the new advanced engineering project. If the expert has already met you in his class and knows your ambition and capability, it is highly likely he will vote in your favor.

The best courses for in-house training are those that teach you how to use the company's resources to get the job done quicker, easier, and, hopefully, with higher quality. Good examples of these courses are those in computer training. They may include courses in such things as training on word processors, E-mail, computer aided design (CAD), computer aided manufacturing (CAM), and computer aided engineering (CAE).

A word of caution about the type of further education you choose. Each type has advantages and disadvantages of which you must be aware. First, all further education is not a guarantee of career advancement. It only becomes a guarantee of advancement when it is combined with excellent job performance. Obtaining further education and performing poorly on the job is of no benefit. Second, further education qualifies you for higher-level job assignments. If the job assignments are not available in your company at the moment you complete your degree, you may be stuck in the same job as before for a long time. Or you may have to change departments or even companies to receive the advancement you deserve. Third, you may appear to the group you are working in as the same old person. They may give you the attitude of "no big deal." This can be very discouraging. You need to recognize it and deal with it appropriately. Finally, your co-workers, may be jealous as you finally complete the degree and management is giving you all this attention. Diplomacy and tact is required on your part when this starts happening.

Returning to night school at a university for a credited degree has its advantages and disadvantages. The degree has the advantage of being recognized by other companies if you should change companies. The degree stays with you every time you change employment. The disadvantage is that you often have to travel great distances and attend classes at night. Returning for a degree or certificate from a university is also a multiyear commitment, which can be difficult. Most people lose interest or have other commitments (i.e., family and children) that intervene.

Satellite courses offer you a chance to attend classes from various universities around the country that you might not normally get the chance to attend. However, taking satellite courses often requires that you take the class during the workday, something your supervisor may not be pleased with. In addition, satellite courses are usually few and far between. They often do not lead to a degree, but still may have benefits for you. Again, evaluate your specific education goals.

Taking after hours courses offered by the company allows you to meet the technical experts in the company on a one-on-one basis. In addition you may meet other co-workers in the company who are probably working problems similar to yours. This is a great way to network in the company. After hours company sponsored courses are not usually accredited or recognized by universities. Therefore they do not lead to a degree and their value quickly diminishes once you leave the company.

For successful career development you will probably use all of the abovementioned forms of further education at some time in your career. The key is to balance the type of education you choose with your career objectives. If you aspire to become CEO of the company, then attending night school at an accredited university for an advanced degree is what you must do. If you wish only to update your technical background, courses offered at symposiums are your best bet. Remember, further education never hurt anyone and it always serves to broaden a person's capability. The broader your background and the more knowledgeable you are the better the chance for career development or job security.

After exploring the opportunities for further education, a majority of people still express some hesitancy. People often use the excuse of "I just don't know." In response to this I remind people of the days when they were attending college, struggling to make it through one class after another. In particular I remind them of the high costs, no medical benefits, no vacation pay, little help from other people, and the uncertainty about getting a job when they finish. I ask them if I could show them a way to go to college and at the same time receive a full salary, paid tuition, full medical benefits, and a paid vacation would they be interested? Not only that, but if they had difficulties with a class someone in the company who had taken the class and would gladly help them out. And when they completed the program they would have a guaranteed job with a high chance of a raise or promotion just for completing it, would they be interested? I have even had supervisors pay for my books, mileage, and parking fees since the class directly related to my work at the time. These are all the benefits a company tuition plan offers.

Now, with all these benefits a company tuition plan offers, would you be interested? After considering all these things, most people respond that they certainly would be interested. Therefore, if you are not getting ahead due to your limited education, don't blame the company—you have only yourself to blame.

In summary, overcoming the educational barriers in your company is a must for career advancement. The best way to hurdle the educational barriers is to use your company's tuition reimbursement plan for further education. Continuing education is a win situation for you and a win situation for your company. The starting place to return for further education is your supervisor or personnel department. You must map out an educational plan that best fits your career objectives. Further education can be obtained through night schools at your local university or community college, symposiums offered by engineering societies, or company sponsored courses. Each type of continuing education has its advantages as well as its disadvantages.

ASSIGNMENT

1. Find out what your company's tuition reimbursement plan covers.
2. Identify someone in the company who has used it and meet with them.
3. What are your long range career objectives? What type of further education is best for you?
4. What societies could you join?
5. What universities are located close by and what courses do they offer?
6. Does your company offer any internal after-hours courses?

CHAPTER 5

USING THE COMPANY'S JOB OPENING SYSTEM TO YOUR BENEFIT

Using your company's job opening system (JOS) to your benefit is one of the most important things you can do for career development. The JOS is the means by which companies inform their employees of the jobs that are available in the company. Are you aware of the JOS in your company and how to use it to your benefit? If not, you could already be missing an opportunity for advancement. In this chapter we will explore how you can benefit from your company's JOS.

Typically, the personnel department in most companies will periodically post all the job openings. The company does this because most times the best candidate for the new position is someone already in the company. The company has to spend less time and effort training a person already working for the company compared to hiring someone from outside the company. So you as an employee often stand a better chance of advancement than an outsider.

Often the job openings are listed in the company newspaper, posted on bulletin boards, or accessible through the company computer network. Do you read these listings at all? It amazes me how many people will read the daily newspaper's sports, financial, or arts section or the comics but hardly ever look at the company's job opening listing. To get ahead you must read the company's job opening listings continuously. You can never tell when a better opportunity will present itself. In addition to keeping abreast of the job market internally you should also be continuously looking outside the company. You should be continuously searching through the trade journals and newspapers for the job listings they may carry. Study and read all the job ads you can. You may even want to contact some professional search firms to do some searching for you.

THE IMPORTANT THINGS YOU CAN LEARN FROM READING THE JOB ADS

Do you need to really study the job ads? Yes! Job ads will provide you with a wealth of knowledge about the job market and your chances for advancement at any given time. By studying the company's job ads you can quickly tell which divisions are hot and hiring. You can also tell which divisions may be having difficulties since they have no openings. How is your division doing? Is it hiring? Chances for promotion are significantly enhanced if you are working for a division that is hiring.

Study the job listings further. Do they identify the job level, type of engineer required, and the pay level? The job ad should identify the experience and background they are looking for. The background sought by the majority of the job ads indicates the background you may need to get ahead in the company. If all the openings are for chemical engineers and you are a mechanical engineer you may have a difficult time advancing. Compare the salaries being offered in the ads to your job level. How does the pay compare to yours? If you are getting underpaid it's time to move on. If the pay is comparable you can be satisfied for the time being.

Determine the salary being offered for the jobs one level above yours. Would you like a raise and a promotion? How will you find out if you qualify unless you go for an interview? Remember, your supervisor may be limited in the salary increase he can offer you. However, it maybe easier for a different supervisor to give you the raise you want simply because you would be a new hire to his department.

Study the job ads to determine which divisions have openings and what products you would be working on. What products are hot and undergoing rapid development? What divisions and products are not growing? This is easy to tell by lack of job ads. Is there any chance of transferring to a new division or product and pick up a promotion in doing so?

A word of caution about company JOS ads. Most companies have internal policies regarding promotions. One of these policies is that if a supervisor intends to promote an employee to a new level, he must have a need and a position open for that grade level. The company wants to put the best person into that position, not just the one the supervisor wants. Therefore they require the supervisor post a JOS listing for the position and interview all possible candidates. The intent is to get the best possible candidate for the position.

What this means is that often a JOS listing will be posted with someone already in mind. Most engineers are unaware of this policy and therefore think they stand a good chance of getting the job. If you are responding to a JOS listing do some research first. Try to find out if there is someone already in mind for the position. The obvious question to ask the supervisor when you are interviewing is "is there anyone presently in your group that is interviewing?" If he answers yes, that person is more than likely the best candidate and it is a "done deal," as they say.

In addition to reading your own company's job listings, you should constantly be checking what your competitors are doing through job listings in trade journals and newspapers. Are they hiring? What type of people are they looking for? What are they willing to pay for employees with a similar background? What are the competitors' growing and expanding divisions and product lines? Are they looking for your background and experience level or some other type? All these questions can be answered through studying the job listings. All this information is very important for career development.

The next question that usually arises is "What do I do if I find something?" The answer is straightforward. Go and check it out. You should check at least once a year for other job opportunities through some type of job interview. This is important for three reasons. First it will keep your interviewing skills sharp and your résumé updated. Second, it may result in a better opportunity. Third, you may find out that your job is not so bad after all. In any case you benefit from the experience and lose nothing.

HANDLING YOUR PRESENT SUPERVISOR WHILE YOU ARE INTERVIEWING

Most people react to the reasons for pursuing career advancement listed above with "That's great, but if my supervisor finds out, I'll lose my job." If this is really the case I highly recommend you make a change immediately. You do not want to be working for this type of person.

If your supervisor does find out that you are exploring new opportunities, there are several things that you can say that should result in benefiting your career. First, you can say that you are happy to be working for him or her, but you heard of this wonderful opportunity and thought you would just check it out. You are happy with your present position and have no intention of leaving. The opportunity sounded interesting and you were going to check to see if it had any substance to it.

Or you might respond with a compliment to your supervisor. You can compliment him by pointing out that the only reason you are looking is that he has done a great job developing you, and the only reason you stand a chance for a better opportunity is due to his excellent training and development!

Still another approach is to be frank and open if you are unhappy with your present position and rate of growth. Tell your supervisor that you do not see much opportunity in his group and you feel that you must explore the options. She can react in one of two ways to this. One way might be by informing you there is not a lot of opportunity in his group for you and you should probably look elsewhere. In this case you have found out some very valuable information about how your supervisor perceives you and what your real chances for advancement are by staying in the group. Therefore you have made the right decision to search for work in other places.

The second way your supervisor might react is to express concern and describe how valuable you are and how he would like you to stay. You have now opened the door to further discussions about your chances for advancement. Use the opportunity to express what your future career objectives are and hopefully the two of you can work out a plan to get you what you want. In other words, use the opportunity to start planning your next promotion together. If you have a good rapport with your supervisor, you might point out that it's easier for him to give you the raise or promotion than hire a new person and spend time training him.

Regardless of what your supervisor says, continue to seek out new job opportunities. If the new opportunity is within your company, make sure you inform your supervisor before you talk to the person doing the hiring. It is professional courtesy to inform your supervisor about the opportunity. Don't let him or her find out from someone else. It will only end up hurting you.

There is a need for discretion when you are looking around. If you decide to go job shopping, keep it a very low profile activity. If you are too visible and start interviewing to many places this can backfire. Your supervisor will quickly find out from other people what you are doing and want to know why. It does not help his career if you are highly visible while you are checking out other opportunities. Do not volunteer any information unless you are asked directly. Don't discuss your plans with co-workers. They will only go running to the boss the minute you leave their office. For this activity, low profile is the way to go.

When you go for the interview, you had better have some good answers to some very difficult questions that are usually asked. The first question is usually, "Does your present supervisor know that you are interviewing?" A very good answer to this question is that your supervisor does know and she does not want to stand in the way if this is really a good opportunity for advancement. The second is, "Why do you want to leave her?" One way to respond to this is to say that you are looking for a better opportunity and you thought this might be one. What ever you do, do not speak negatively about your present position or employer. Anything that you say about your supervisor or group will immediately be spread all over the company. Do not open Pandora's box.

Remember that soon after you leave the interview the first thing the hiring supervisor is going to do is call your present supervisor. And how your present supervisor responds can immediately make or break the opportunity. If you are on good terms with your present supervisor and she thinks highly of you, she may respond with "He is a great employee and I don't want to lose him." If your supervisor says this, it will probably make the hiring supervisor want you all the more. If she responds with "He is a marginal performer and I'd like to get rid of him" this can end the opportunity immediately. The point here is "Do you know how your supervisor will respond?" If you don't you are taking a big risk. You should know how your supervisor is going to react before you spend time interviewing. Therefore, keeping on good terms with him during this whole process is essential.

If you are already on poor terms with your supervisor this can be extremely good or poor for you. Sometimes I've seen supervisors go out of their way to help an employee get a new position when they are on poor terms. The supervisor wants to get rid of the employee. So rather than fire the employee, the supervisor finds new opportunities for the employee and may even pass along great recommendations about the employee's per-formance. The hiring supervisor receives a glowing report about the applicant's past performance so she will be encouraged to hire him.

Other times I've seen the supervisor take a different approach—they use the opportunity to get even. Due to the poor relations, the supervisor will purposely underrate the employee's past performance and spoil any opportunity for advancement. If you have poor relations with your supervisor, you might want to ask him how he would react before you go through all the effort. Watch his reaction. If he is giving you the "get even" response you will have to deal with it. You cannot ignore it.

HOW TO GET THE MOST FROM A JOB OFFER

After you get a job offer your work is still not done, regardless of whether the offer comes from inside or outside the company. In either case the ideal situation is to get your present supervisor to make a counter offer in the hopes of keeping you. You should inform your supervisor of the new offer. Explain why it appeals to you and why you intend on leaving if a counter offer is not provided. The intent here is to try to get an even better offer out of your present supervisor.

It is surprising how fast some supervisors can move and what they can counter offer when they know they are about to lose one of their key performers. Ask him to at least match it. Sometimes another offer is just what is needed for building a case for promoting you with your supervisor. If someone thinks you promotable, why shouldn't your present supervisor think so too. Sometimes this is all that is needed to push some supervisors into action. It hurts their ego to know that someone else in the company is trying to steal away their people.

If you get your supervisor to counter the offer, your efforts are still not finished. Take that offer to the new supervisor and find out if she can offer more. Your goal is to get both supervisors to offer the absolute most they can. You repeat this process until both supervisors can no longer offer anymore. At this point you win!

Some engineers feel this unethical or not normal. To this I say look at other professions, athletes in particular. Athletes even go as far as hiring agents to negotiate job opportunities for them. Actors, actresses, lawyers, and nearly every other profession do the same thing. Knowing how to negotiate is essential for career advancement.

WHICH IS BETTER, INTERNAL OR EXTERNAL OFFERS?

When the best and final offers are in, you need to consider other aspects before you make a job change. If you change jobs for a promotion and stay within the company this is the best of all possible worlds. By staying within the company you keep your vacation benefits, seniority, and retirement benefits. Studies have shown that an engineer who stays with one company until retirement, in most cases, will retire better than someone who has changed jobs. This is the case even if the person changing companies has had larger pay raises.

Remember, when you leave a company you lose your vacation benefits and retirement benefits. Leaving the company may also cause you unexpected expenses you never planned on. For instance, if you have to drive further to work or move out of state there are additional expenses that you will incur. There will be costs to change your car insurance, place of residence, driver's license, and real estate fees just to name a few. Do you have any of the raise left after you pay all the new expenses? How will your retirement benefits be affected? For example, a 3% raise inside the company may be equivalent to a 6% increase outside the company.

Another point one must consider is that in the new company you will be starting all over. You can not expect a large raise any time soon. It will take you over a year to reestablish your position and your performance. If you get a raise when leaving make sure it's large enough because it may be a long time before you get another one.

> Remember, there is more than a simple pay raise involved when one considers leaving the company. You will have to decide for yourself whether the move is worth it or not.

If you are lucky enough to be offered a new opportunity in or outside the company, make sure you give your supervisor at least a chance to match it if you still would like to continue working in the same group.

WHAT TO DO IF YOUR SUPERVISOR TRIES TO BLOCK YOUR MOVE

If your present supervisor tries to block your move you have several options. First try to get your supervisor to agree to a convenient transfer date some time in the future. Let your present boss know that you are not going to leave the company high and dry. You should be able to work out some gradual transfer plan. If supervisor continues to fight you, you will have to play hardball. What you can do is point out what your reaction will probably be over time. At first, you will continue to work hard but after a while you will probably lose interest. It's hard for you to give it all you've got when you gave up a promotion and raise simply because your supervisor didn't think you should get it. Try to make him see that holding you back is not good for him, for you, or for the company in the long run. Nobody wins in this type of situation. Another thing you might try is to volunteer your help in finding and training your replacement. This always makes the transition easier.

ALWAYS LEAVE ON GOOD TERMS

If you do accept a raise and promotion from elsewhere in the company there are several things you must do. First, make sure you thank your present supervisor for all past help. Point out all the good things your new supervisor liked and how you qualified for the new position in part due to the fine effort your old supervisor did in developing you. Never leave on bad terms! It will only come back to haunt you later.

In summary, using your company's JOS to your benefit is important for your career development. The JOS allows you to compare what you are making to others in the company. The JOS is the first place new jobs are usually listed, and by constantly reading the ads you will be aware of the latest opportunities. In addition to reading your company's JOS you should be reading the trade journals for job ads. These ads will tell you what people outside the company are making as well as give you insight into what the competition is doing. When you consider taking jobs outside the company take into account all the new expenses you will incur and all the benefits you may lose. And, finally, if you decide to make a move, leave on good terms, do not burn your bridges behind you.

ASSIGNMENT

1. Find out how your company JOS works.
2. Study the ads. Do you know all the codes? What do the ads tell you? Which divisions are hot, which are not?
3. Review some trade journals to see if you can find the job ads. Study them. What are they looking for and how much are they willing to pay?
4. Pick one JOS listing and go on an interview.

CHAPTER 6

DETERMINING THE FORMAL AND INFORMAL CRITERIA BY WHICH YOU ARE JUDGED

Do you realize that your supervisor and her superiors are evaluating your performance every day? Do you know the criteria they are using to rate your performance? Do you know what they consider important and what is not? If you have not taken the time to find out what they consider important you could be wasting a lot of time and energy. Like a blindfolded child, you swing wildly, trying to hit the piñata. That child has no idea where to swing but hopes that with the next swing he or she will hit the piñata, break it open, and retrieve the prizes.

If you don't know the formal or informal criteria by which you are being judged, you are acting like a blindfolded child. You blindly perform task after task hoping the next one will get you promoted. Like that child, you blindly swing at the piñata or promotion, which you cannot see. Your hope is that everything you do will result in getting the big raise or promotion. Just remember it is very difficult to hit a target that cannot be seen. And blindfolded you can spend a lot of unnecessary time and energy chasing the promotion. Often when the child succeeds in smashing the piñata it is due to blind luck.

Are you operating with a blindfold on as far as your career advancement is concerned?

You are if you do not know the formal and informal criteria by which you are judged.

Remove it by discovering the informal and formal criteria by which you are judged!

Don't leave your next promotion up to blind luck. Take the blindfold off.

In this chapter we will discuss the methods you can use to find out the exact formal and informal criteria being used to make judgments on performance. This will in effect remove the blindfold. Once you have clearly identified the criteria for making certain judgments, you will be able to see exactly what you must do to get promoted. With the criteria clearly identified, it is easier to get the raise or promotion. It should also come sooner since you can make every work task you perform help you fulfill the criteria.

FORMAL CRITERIA: WHAT ARE THEY COMPRISED OF?

In every company there are formal criteria by which you are judged. There are also informal criteria. You must understand both criteria for successful career development. The formal criteria are usually well defined and documented.

The formal criteria are manifested in three ways. The first is through the *job performance review process*. The second is through the *job performance criteria or guidelines*. These guidelines summarize the performance that is expected of the employee at each level of the engineering ladder. The third way the formal criteria are manifested is through the *promotion review process*. Each company has its own promotion review process that all supervisors must follow to ensure employee promotion approval. This process is usually a very formal one that is well defined and strictly adhered to. We will now explore how you can clearly identify and deal with each of the three formal criteria.

UNDERSTANDING THE JOB PERFORMANCE REVIEW PROCESS

The first formal criterion with which you must become thoroughly familiar is the job performance review process or job appraisal method utilized by your company. On a periodic basis your supervisor must formally document your performance on the job. Some companies review employee job performance once a year or once every other year. You must find out everything involved in the job performance review process for your company. How often do they conduct job reviews? When is your next review? What criteria are involved? A good way to clearly identify the criteria is through the paperwork that is processed during your job review.

In preparation for a job performance review, a supervisor will fill out some type of form that documents formally your performance to date. Some companies use a standardized form and other companies simply document your progress in a memo. Ask your supervisor for a copy of the form used if you have not yet had a job review or performance appraisal. Study the form and make sure you are familiar with everything on the form. A sample employee job performance review form is shown in Figure 6-1. Keep in mind as we review this form the things you will need to look for in the form utilized by your company.

Across the top of the form is usually the employee data. This data includes the employee name, job title, grade level, years in grade or job, department, and some reference to the last appraisal date. Review your form and see what data the company considers important. Why do they consider it important? For example, some companies have a standing

PERFORMANCE COMMUNICATION PROGRAM			
RATING FORM			
EMPLOYEE NAME & JOB TITLE	EMPLOYEE SS NUMBER	GRADE	YEARS/MONTHS IN GRADE
DIVISION/DEPARTMENT	APPRAISAL DATE	PERFORMANCE PERIOD	FROM TO

WHEN COMPLETING THIS FORM REFERENCE THE BASELINE AGREEMENT AND DISCUSSION QUESTIONS

PERFORMANCE CHANGE RATING

[ ] 1 GROWING RAPIDLY IN GRADE - EMPLOYEE IS DEMONSTRATING RAPID IMPROVEMENT IN PERFORMANCE & CAPABILITIES TO HIS/HER PAST PERFORMANCE & OTHERS IN GRADE.

[ ] 2 PROGRESSING WITH/GROWING IN GRADE - EMPLOYEE IS DEMONSTRATING IMPROVEMENT IN PERFORMANCE COMPARED TO PAST AND OTHERS IN GRADE.

[ ] 3 PROGRESS LESS/ DECLINING IN GRADE - EMPLOYEE PERFORMANCE IS UNCHANGED WITH NO GROWTH OR HAS DECLINED COMPARED TO PAST & OTHERS IN GRADE.

JOB REQUIREMENTS RATING

OVERALL RATING *

[ ] 1 CONSISTENTLY EXCEEDS JOB REQUIREMENTS

[ ] 2 MEETS/OCCASIONALLY EXCEEDS JOB REQUIREMENTS

_____

* (PERFORMANCE CHANGE + JOB REQUIREMENTS) / 2

[ ] 3 DOES NOT MEET JOB REQUIRMENTS

CAREER DEVELOPMENT

I WANT A CAREER DEVELOPMENT PROGRAM DISCUSSION WITH MY SUPERVISOR [ ] YES [ ] NO

REMARKS, SUMMARY, STATEMENTS, ETC.

SIGNATURE INDICATES A PERFORMANCE COMMUNICATION DISCUSSION BETWEEN EMPLOYEE & SUPERIOR HAS OCCURED

EMPLOYEE	SUPERVISOR	MANAGER

DISTRIBUTION: WHITE - HUMAN RESOURCES   YELLOW - EMPLOYEE   PINK - DEPARTMENT FILE

FIGURE 6-1   Sample employee job performance review form.

policy that an employee must be in a grade for a minimum of three years before they can be promoted. This is the reason for the time-in-grade block on the form. If your company has a similar policy, what do you think your chances are for a promotion if you have only been in the grade one year? Does your company have any other hidden prerequisites for promotion of which you need to be aware? For this reason you must understand the importance of all the data on the form.

Below the general information employee data is the performance summary section. *This is the bottom line.* All forms have one. This section is where in a matter of two or three lines your entire performance is summarized. Most people do not realize the importance of this section. This section usually summarizes your performance and identifies some type of overall rating. Your raise is usually computed based on these ratings. Another way you can think of this section is in terms of dollars. For the example shown, a 3 rating results in a demotion or salary reduction. In other words, it may cost you thousands of dollars and years of setback. A 2 rating means status quo. Exhibit just average performance and you'll receive the average raise and, in thirty or forty years, the next promotion. Finally, a 1 means outstanding performance—you've earned a large raise and, by all means, keep up the good work; you will soon be experiencing that promotion. After identifying the hidden meanings behind these formal performance rating criteria, most people immediately start to pay more attention to this section.

Do you understand the section of the job performance form that summarizes your rating? How do the ratings get translated into dollars? What do they mean in terms of promotions? If you don't know, you are operating with a blindfold on. Ask your supervisor; he or she will tell you how the rating translates to raises and promotions.

Below the performance summary section is the career development section. This section identifies your desire to discuss a career development program with your supervisor. As far as I'm concerned there are only two reasons why you would not want to check yes in this section. One reason is that you are making more money than you think you should and therefore you don't feel that you need career development. The other reason is that you plan on retiring next year and therefore career development is not uppermost in your mind. If you do not fit into either of these categories than you should be requesting a career development discussion with your supervisor.

If your company's form does not have a section similar to this, ask your supervisor for extra time to discuss your career plans. You may want to schedule this talk when you both have more time to discuss your future plans. Do not try to get through your performance review and have a career development meeting at the same time. It is too much to try to cover in one meeting.

The next section on the form is the remarks and summary section. This section is usually filled out by your supervisor. In this section your supervisor will try to summarize your performance since the last performance appraisal. Study the remarks carefully. On what matters has comment been made? What has not been noted? Do you know why things were missed? What did he consider important? What did he consider unimportant? After you answer these questions you may come to realize what your supervisor considered important was not at all what you considered important.

At some point in the performance review you will get to the "Needs Improvement" part. In this part the supervisor informs you what things you need to improve upon. Make sure you spend enough time on this subject. It may be hard to sit there and listen to your supervisor spell out all the things you need to improve upon, but it is a must for career development. By identifying your shortcomings, your supervisor is telling you the very things you need to improve upon to get the raise or promotion.

Whatever you do, please, don't take the criticism personally. Don't try to find an opening in the commentary wherein you rebut anything you interpret as detrimental. You are only going to lose. Your best bet is to listen patiently to every criticism and make sure you understand it. From this criticism you can learn what he or she considers important that you are not doing, the weak areas that are holding you back. If the supervisor lowers both guns at you, you need to consider whether you really want to continue working for him or her, or if perhaps it's time you moved on. Most performance appraisals do not go this badly even though they may seem that way. Listening to the criticism is tough. Everyone has a difficult time doing this. Here are some tips on how to turn this difficult time to your advantage.

One way to turn this difficult time to your advantage is to use a tag-on statement after each criticism. Typical tag-on statements go something like this.

I understand now, so if I improve my performance doing . . . (*name the criticism your supervisor just identified*) . . . then do I stand a better chance for a raise?

Let me make sure I understand this better. One of the reasons that I did not get the promotion was because . . . (*name the criticism your supervisor just identified*) . . . and if I correct or improve myself there should be no reason why I would not get the promotion next time?

Let's summarize. The areas that I need to show improvement on in order to get the promotion by the next appraisal time are . . . (*name your improvement areas*) . . . and if I improve these things I stand a good chance for advancement?

Are these the only things that I need to improve upon to qualify for the promotion by the next appraisal time?

By adding these tag-on statements to the criticism you are doing three things. First, you are clearly identifying what has caused you to miss the big raise or the promotion this appraisal time (removing the blindfold). Second, you are hopefully getting your supervisor to identify everything you need to do to get the promotion next time (identifying any hidden agendas your supervisor may have). And, third, you are sending the message that you expect the promotion by the next appraisal time since you will be improving your performance in each of the identified areas (setting the deadline for the raise or promotion). You have clearly sent the message to your supervisor that you understand your weaknesses. You will correct the problems and after doing so you expect the raise or promotion next time. Remember, people rarely get promotions they do not ask for.

Another tip is to think of each criticism as a step closer to the next big raise or promotion. This should help to make acceptance of the criticism much easier. Every criticism that your supervisor identifies becomes another reason why you deserve the promotion after you have proven to your supervisor that you corrected the problem. The more reasons your supervisor identifies, the more reasons you will have in your defense once you have overcome the problems. When your supervisor does not identify areas for improvement then it is time to worry. If this happens, you need to get him or her to open up more.

My own experience has shown that the quickest and best promotions came only after I got my supervisor to really open up and clearly identify what I had to improve. It was only after we got everything out in the open and I started to improve my performance that she soon realized there were no reasons not to promote me and she quickly got started getting the promotion approved.

At the end of your performance review you should have a clear understanding of what you need to do from what is recorded on the appraisal form. From this point on, getting the raise or promotion becomes much easier. As you improve in each area, simply point out your improvement to your supervisor. Make sure you highlight how you are meeting his or her criteria to qualify for your next raise or promotion. It also does not hurt to point out the same things to your manager as well.

This brings us to the last section on the appraisal form, the signature block. Who has to sign it? The signature block tells you immediately who controls your raises and promotions. Do you know who will sign your appraisal and approve it? If you do not know them, then they surely do not know you. And most supervisors or managers do not promote people they do not know. Some people do not even know who will see their appraisal form. Meeting with and getting to know the people who sign your appraisal form is the key to career development.

The sample form in Figure 6-1 may not look at all like your performance appraisal form. I have used it as a guide to help you study your form(s). Some companies use appraisal forms that are several pages long. Some forms require the supervisor to rate you in each area of your job that the company considers important. Whatever type of form your company uses, make sure you are familiar with everything on the form. Understand the hidden meanings behind each block. Only after you have taken the time to understand the formal criteria defined on the performance appraisal form can you expect to use it as a guide for your career development.

DETERMINING HOW JOB PERFORMANCE IS MEASURED IN YOUR COMPANY

The second formal criterion you must be aware of is the one your company requires you to meet at each level on the engineering ladder. The starting point to discovering this criterion is your supervisor. Your supervisor has written guidelines that define the performance expected at each level of engineering. Meet with your supervisor and ask him for a copy of the criteria. Usually these criteria have been developed over time and with the help of the personnel department. If your supervisor does not have a copy, stop by the personnel department and request a copy. They should be very willing to help you out. Make sure that you obtain the formal description of the level you are at and the next level you hope to reach when promoted.

Before asking your supervisor alot of questions about the guidelines, study them. Try to identify all the things you have accomplished at your level and the things you need to work on. Next look at the level above yours and study it to find out what you have to do to reach that level. In studying the guidelines you will probably generate more questions than answers. This is good.

Since every company will have its own guidelines Figure 6.2 is a sample guideline based on the guidelines of several companies. Studying this guideline should help you in understanding yours. This sample guideline is organized in a matrix fashion with the job grade levels defined in the first column and the criteria categories defined across the top. For this example five levels of engineering are defined and six performance categories are identified for each job level. The job levels are ranked from entry-level Grade Level I engineering to the most senior Grade Level V. The six performance categories contain two groups of skills. The left three columns identify the technical skills needed by the engineer and the right three columns identify the interpersonal or team leadership skills needed.

First let me try to show you the differences between the engineering levels with an example of how the job responsibilities vary for each level. At engineering level I, you might receive an assignment that requires you to analyze some data. The data have all been collected, the analysis has been defined and programmed into a computer, and the program output is on a graphics plotter. Your job is to enter the data and generate the plots with the existing program and computer.

At grade II your assignment will be to collect the data from the test, enter it into the computer, modify the program if necessary, and plot the results.

At level III your assignment will be to collect the data. But first you have to assemble the hardware, plan tests, and obtain the help of a technician to collect data, organize the results for input into the computer, and plot them out with the help of junior engineers.

At level IV your assignment will be to collect data and analyze it. But first you must organize a team, get a time and cost estimate to do it, schedule tasks and make assignments to collect the data, write a program to analyze the data, choose a computer to complete the task on, identify type of language, and define the plots to be generated. You will do this by organizing a team of engineers, computer programmers, and technicians.

At level V your assignment will be to figure out what has to be done. What tests are to be done? What data is to be collected? What does theory predict? But first you must organize a team, get cost estimates established, brief management on plans, and get approval. Next schedule tasks and make assignments to accomplish your experiments and collect data. You must oversee the team's choice of computers and language as well as

GRADE LEVEL	TECHNICAL REQUIREMENTS	TECHNICAL JUDGEMENT	TECHNICAL CHALLENGE	LEADERSHIP & WORK DIRECTION	MANAGMENT OF COST & SCHEDULE	INTERACTION
I	Supportive of Project Engr. 4 to 5 Yrs Engr. Training	Evaluate & Recommend Technical Solutions	Applies Known Techniques	Can Explain & Coordinate Work with Individuals of Same Grade	Performs Assigned Tasks within Specified Cost & Schedule	Custmer Contact Not Normal Coordinates with Own Group
II	Perform Basic Engr. Tasks Perform Analytical Prediction of Results	Evaluate & Recommend Effective Solutions Justify Solutions Based on Facts	Technical Precedent Usually Defined Defines Tasks to Be Performed	Can Explain & Coordinate Work with Individuals of Same or Lower Grade	Performs Required Tasks within Cost & Schedule Estimate Supporting Goods & Services Required	Coordinates with Other Departments & Projects Infrequent Customer Contact
III	Experienced Performer in a Speciality Capable of Performing in Broad Range of Assignments	Evaluate Alternatives & Select Technical Approach Justify Alternative Selected	Can Identify Tech., Cost, & Schedule Constraints Can Apply Original Approaches Based on Established Precedents	Can Direct a Small or Specialized Team in Pursuit of a Task Objective	Accomplish Task/Team Objectives within Cost & Schedule Anticipates Problems and Initiates Action	Regular Supporting Role in Customer Contacts Usually Limited Technical Exchange
IV	Experienced Leader in His/Her Field Has Depth of Knowledge in Related Fields Demonstrated Ability	Select & Implement State-of-the-Art Solutions Precedence May or May Not Exist	Technical Precedence Not Established Original or Creative Approach May Not be Required Makes Tech., Cost, & Sched. Judgements	Can Organize & Direct a Small or Specialized Team in Pursuit of Project Objectives	Accomplish Project & Team Objectives within Cost & Schedule Cost & Schedule Implications Typically Have Significant Project Impact	Regular Tech./ External Contact with Customers Normally Has a Limited Role for Tech. Interface Communication & Judgement Key to Success
V	Recognized Authority Grad Work or Adv. Work in Field Advice Solicited by Mgmt. & Customers	Technical Opinions are Respected Internally and by Customer Technical Decisions Typically Reviewed by Results Only	Technical Solutions Beyond Industry Precedents Assimilates Complex Problems & Interactions Has Tech. Solutions	Can Define the Need for & Direct the Work of a Group Concerned with a Variety of Engineering Disciplines	Accomplish Project/ Team Objectives within Cost & Schedule Select Alternatives Involving Cost, Sched., & Tech Trade-offs	Regular Ext. Customer Contact Lead Role in Customer Interface & Tech. Solutions Key Individual to Ensure Project Success

FIGURE 6-2 Typical engineering grade level performance guideline.

plotters. Once the data have been collected you must get the team to write a final report and you must present the results of your effort to management. If technical problems arise you are expected to determine the best methods to modify the experiments and resolve the issues.

This simple example shows how to identify the differences between grade levels. The example also shows that as you move up the chain interpersonal skills and team leadership skills become more important. Supervisors are not really expecting much in the way of team leadership skills from junior engineers. However, you should be developing them as you go. Management is expecting good leadership skills from the upper grade level engineers. Therefore the criteria on the right of the chart become more and more important the higher you advance.

All supervisors interpret these generalized guidelines differently. Some supervisors firmly believe that all you have to do is be good on the left side of the chart. Other supervisors believe the right side of the chart is most important. All supervisors think differently. On your company's guidelines, do you know what your supervisor considers the most important? You could be working to impress your supervisor with your technical judgment skills and she may be thinking that leadership is the most important.

Another common mistake that engineers make when looking at the guidelines is interpreting the level one should be rated at. Your job assignment may allow you to perform at many levels. Let's assume an engineer has the following ratings for each performance category:

Technical Requirements	Level III
Technical Judgment	Level III
Technical Challenge	Level IV
Leadership and Work Direction	Level IV
Cost and Schedule	Level V
Interaction	Level V

What level would you rate this engineer? Most people respond with level IV. However, from a management point of view this is wrong. The engineer is rated on his lowest level of performance. To be rated a level IV the engineer must demonstrate performance at level IV in all categories before she will be considered ready for promotion to that level. This level III engineer has a good start on promotion to level IV but some areas must be worked on before promotion.

Why have I put these guidelines into this book? Not for you to study them and learn about engineering levels. Not for you to determine how you are performing. I did it because it's a cheat sheet! Its purpose is for you to use it that way. Here is how you do it. Copy down these guidelines on a single sheet of paper and make sure they are readable. Now take them into your supervisor. Explain that you got them out of a career development book and sit down with your supervisor and ask for an opinion. How important are these criteria and what does your supervisor see as the real differences between grade levels. From that point on don't talk; listen to what is said. The guidelines are too general for anyone to really determine the criteria for each grade level. Therefore, anything said defines the criteria as your supervisor sees it.

Everything he or she will be telling you will be exactly what you will need to know about his criteria for development. Have him explain what he considers the most important criteria for your next level. How does he rate you for each category and where do you need to improve? Remember, you should be all ears at this point because he will now be telling you everything he considers important. Try and absorb as much as you can. You should be like a sponge soaking it all up. As he is talking, make notes all over the guidelines. Mark them up together, cross out and change things to his satisfaction. When you are done you will now have everything on one sheet of paper that identifies exactly what your supervisor thinks and what you must do to earn the next promotion. In effect you will have a piece of paper that has most all the answers on it—a perfect cheat sheet to help you shorten the time to your next promotion.

If the sample job level guidelines are too different from your company's guidelines, then use yours. Meet with your supervisor on an informal basis and review the company's job grade guidelines together. I recommend that the best place to do this is at your desk or office, where it is not intimidating to you or your supervisor. Your supervisor's office can sometimes be too intimidating. You also stand a better chance of not being interrupted by other people and phones.

Ask your supervisor to explain the performance expected by the company at your level. Then ask her to explain the performance it would take to be at a level up from yours. Once you get him or her talking, don't interrupt. You should be all ears at this point. Remember, you have two ears and one mouth; you should be doing more listening than talking. Whatever happens, do not start arguing about the interpretation of the guidelines. Make notes in the side margins for everything you can. If you do this right, at the end of the conversation you will have identified the formal criteria that she considers you must meet for the level you are at and the next level above yours.

During your talk try to get your supervisor to identify some specific things you have to demonstrate in your work that will show you are meeting the criteria for your job and possibly for the next level. For instance, you might ask, "If I complete my assignments on time and within cost, does that signify performance at my level or the one above my level? How might I perform my tasks such that I will have demonstrated I meet the criteria for the next level? What exactly do you see as the difference between my level and the next one above? How do I get broader assignments so I can demonstrate performance above my level? What exactly need I do on my present assignment to increase my chances for a raise or promotion?"

Your supervisor's response to your questions will help lift off the blinders and provide guidance for your career advancement. You now have a clear vision of the formal criteria you must meet for career advancement.

The guidelines are usually written in such general terms that anything defined specifically is really what your supervisor thinks you need to do. Ask as many open-ended questions about the guidelines as you can. The reason for this is that any answer given you will define the criteria. He will be describing exactly what he believes you have to do to get the raise or promotion.

Most supervisors will not feel apprehensive about discussing this with you in a relaxed, nonformal setting. If you wait until job review to have this conversation, it will be too late and too formal. Your supervisor may be as nervous as you are in a formal job review and may feel that you are trying to second guess her. You must have this conversation well in advance of your job review. This way your supervisor will not feel like she is being put on the spot and you will have enough time to demonstrate the performance needed before the formal job review. In any case, make sure you listen and ask her to clarify anything you may not understand. The more she talks, the more you win and the clearer the picture you have of the formal criteria you must meet. Take as many notes as you can and be sure to review them often.

Most engineers believe that the minute they are performing at a level above theirs the company will immediately promote them. It is very discouraging for them to find out that this is not the case. The engineer must go through a formal promotion review process and be judged ready for the promotion. In larger corporations this formal promotion review process often takes months and may involve a multitude of other people. Therefore, the third group of formal criteria that you must be aware of in your company is the promotion review process.

UNDERSTANDING THE PROMOTION REVIEW PROCESS AND USING IT TO YOUR ADVANTAGE

The formal promotion review process identifies the steps that your supervisor must go through to get your promotion approved. For the lower levels on the engineering ladder, this process may just involve your supervisor and his superior. As you move up the engineering ladder, the promotion process becomes more complex and may include promotion review boards made of several people as well as the use of company totems. For career advancement you must have a clear understanding of the promotion review process for your company. Not to know the formal promotion review process is like being the blindfolded child swinging wildly trying to hit the piñata.

To help you better understand the promotion review processes companies go through I will describe two sample processes that represent what most companies may follow for promoting people. The processes I will be describing are generalized in nature but show the dynamics that may be occurring at your company. This should help you in identifying and understanding your company's process. Remember, your challenge is to define this process for your company and your supervisor.

To aid in this discussion I have diagrammed the spheres of influence often involved in the promotion review process. Figure 6-3 shows the spheres of influence that you need to be aware of in the promotion review process. The sphere at the bottom is you. The next highest on the ladder is your immediate lead engineer. Above the lead engineer is your supervisor. Above your supervisor is her superior or manager as well as the company totem and promotion review board.

I have drawn the spheres in relation to how you most likely perceive them. Your lead engineer has the responsibility for assigning the daily tasks and doing most of the interfacing with you. Her sphere blocks most of your vision of the ones above. Her responsibility is to take care of most of the problems. Your supervisor simply has too many people reporting to her to spend the amount of time that she should with everyone. Therefore your supervisor calls on the help of the lead engineers to hand out assignments and make sure things are being accomplished.

Behind your supervisor are three spheres that you need to be aware of. Usually these spheres are not very visible to the engineer. However, knowledge of their existence is a must for career advancement. The first sphere is your supervisor's superior or manager. The manager must approve all the raises and promotions your supervisor requests. In addition

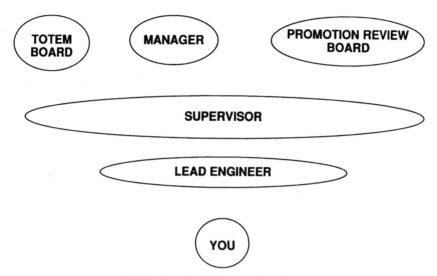

FIGURE 6-3   Spheres of influence.

to your manager there are two other spheres: the totem board and the promotion review board.

The totem board is usually comprised of second level managers and possibly directors. The purpose of the totem board is to compare and rank the performance of all the engineers in the company. The managers meet and rank the performance of all engineers relative to one another. You can think of this as similar to class ranking. Totems are often done once or twice a year. Usually the top-ranked individuals in the totem are the first to be promoted.

The promotion review board is usually made up of a cross section of people in the company. The promotion review board may contain senior staff engineers and upper level managers and personnel. The purpose of the promotion review board is to review each candidate for promotion and determine if they fulfill the criteria established. The promotion review board is a nonbiased third-party group whose function is to ensure that each candidate meets the criteria. Usually your supervisor and manager present to the review board on your behalf the reasons why you deserve the promotion. The review board evaluates the merits of your case and makes a decision based on the evidence presented.

To understand all the dynamics involved in getting a promotion let's first review an example of a lower level promotion where only your supervisor and manager are involved. Generally, for a lower level promotion only your supervisor and his manager need to decide that you are ready. The first thing your supervisor does is to check with the lead engineer about your performance. How do you get along with your lead engineer? Do you support management and follow management's direction? Or are you on poor terms with management and constantly in disagreement? What do you think management's recommendation will be about you? Correspondingly, do you have enough visibility so that your boss can see that you are ready for a promotion or does the lead engineer do all the interfacing. This is a delicate situation that you must keep in balance. You must follow the direction provided by your lead engineer but you must also get visibility with your supervisor. Going around your lead engineer and always directly interfacing with your supervisor can cause problems. You can hurt your chances for a promotion by doing this.

Try to maintain a balance in this situation. Make sure that your lead engineer is always aware of what you are doing. If you have to discuss things directly with your supervisor and leave the lead engineer out, then make sure he knows why you are doing this. On the other hand if your lead engineer is not giving you the chance to report your progress to your supervisor, discuss the situation with the lead. Try and get him to allow you to report progress at least once in a while. Every lead engineer and supervisor differs in how much interfacing they allow the junior engineers to have. You will have to strike a balance with each one.

Once the supervisor and the lead engineer decide you are ready for a promotion, then comes the job of convincing the supervisor's manager. The supervisor usually carries forward a recommendation for your promotion. You must keep in mind that the manager may have anywhere from 40 to 50 engineers reporting to her. And at any given time four or five of those engineers may also be up for promotion. What will make the manager approve your promotion over the others? This is the question or problem that you must address. It is hard enough getting visibility with your supervisor let alone getting visibility with the superior.

The dynamics involved in obtaining the approval for your promotion from the manager will depend upon several things. First, does your performance warrant a promotion? Second, what is the relation between your supervisor and his superior. If they have a poor relationship, it could be years before you are promoted. Third, who else is up for a promotion? Often a limited number of promotions are allowed at any given time due to budget constraints. Is this the first time you have been nominated? Are there others who have been waiting longer? And, finally, how well does your manager know you? If you have never met, you can be pretty well assured that he does not know you and, consequently, unless your supervisor does a wonderful selling effort, you most likely will not get promoted.

Don't panic, there are ways to get around all these roadblocks. I will address them later in the book.

Now let's look at the dynamics involved in a higher level promotion where your manager must utilize the totem board results and present your case to the promotion review board. What happens here is that the manager, with the help of your supervisor, prepares a small brief or summary as to why you deserve the promotion. The brief contains a description of your accomplishments and shows how you meet the criteria for promotion. In addition to the brief, your ranking on the totem is reviewed. This is a check and balance process. If you meet the criteria then you should be performing at the top of your grade and ranked accordingly. If you are ranked at the top of your grade then your performance should normally meet the criteria for promotion as defined by the promotion review board. If you're not at the top of your rank or do not meet all the criteria then your chances are not good for getting the promotion approved.

Once you understand this process you quickly realize you must know two things. The first is where you stand on the totem and the second is what are the board's criteria for promotion. Finding both these out may seem impossible, but it's doable. Finding out where you rank in the totem or in relation to others takes tact and patience. Your supervisor knows where you rank in the totem but does not like to tell anyone. Therefore you may have to do a little fishing. Find out when the last totem ranking occurred and when the next is scheduled. If your company does not use a totem then you must explore how the company ranks engineers. Your supervisor should have this information. Most supervisors will share with you how and when it occurs. The next thing you must find out is where you stand on the totem ranking or in relation to others being considered for promotion. The best approach is simply to ask. If like most supervisors, you will be informed this information cannot be shared. You might find out something by asking what quarter you are ranked in—the top, upper middle, lower middle, or bottom. Your supervisor may share this information with you since he or she is not telling you the exact rank. If you are in the top quarter you're in great shape. If you're below the top quarter, you have your work cut out for you.

The second thing you need to find out is the criteria used by the promotion review board to determine if someone is ready for a promotion. Sometimes this information is closely guarded and sometimes it is readily available. The first place to start asking is your supervisor. If your supervisor has the information and shares it with you, you are in great shape. Study it and determine where you failed to meet the criteria. Once you have done this you will have a clear understanding what you need to do next. If your supervisor does not have the information try to see if her superior or your manager has a copy of the criteria. Schedule a meeting with your manager and see if she will share with you what she thinks you need to do to meet the criteria.

The next place to check is with someone who was just promoted to the level which you are looking to get promoted to. Chances are he or she did not make it on the first try and needed to do some additional things to meet the criteria. By talking to everyone you can who recently received a promotion you can get a pretty good idea what the promotion review board criteria are or what is involved in the promotion review process. If you do not know of anyone, check with personnel and find out who in the company was just promoted. People who were just promoted are usually glad to share this information with you. They know how hard it can be to get promoted. Another good place to look for people who were recently promoted is in the company newspaper.

The important point here is to check all the available sources you can about the promotion review process. The promotion review process will vary from company to company and supervisor to supervisor. It is a waste of energy and time to go after the promotion blindfolded. You can literally remove part of the blindfold by doing research on the promotion process used in your company. Your challenge is to find out all you can about the promotion process in your company and what the formal criteria are that you will be judged against. The information is all there. It's up to you to do the leg work and find it.

This concludes the discussion on the formal criteria. We will now shift to the informal criteria. Often the informal criteria by which you are judged contribute more to determining your readiness for a raise or promotion than the formal criteria. The informal criteria are not written down and vary from supervisor to supervisor. Let's explore some of the more common types of informal criteria.

INFORMAL CRITERIA—WHAT THEY ARE, HOW TO IDENTIFY THEM, AND HOW TO USE THEM

Most engineers do not realize that informal criteria often play a bigger role than imagined in evaluating one's performance. Informal criteria are very hard to define and therefore you must be aware of the subtle hints that are available to you in identifying them. The informal criteria are those intangible things that your supervisor and his superiors use to judge your performance. Informal criteria are usually based on personal biases or beliefs as to what performance is needed. These beliefs are often a result of your supervisor's work experience as well as his or her background and training. The informal criteria are usually never documented and very hard to determine. Some of the informal criteria that I have observed are described below.

Neatness or Appearance

Take a good look at your supervisor and his superior. How do they dress and wear their hair? How do you dress and wear your hair? Some supervisors like to come to work every day wearing a suit. They have neat, short hair. If you believe in dressing very casually and wear your hair long, how do you think your superiors perceive you? Do other supervisors like to come to work casually dressed? How does your style of dressing compare to your supervisors?

Neatness of Office

Take a few minutes and observe your supervisor's office. Some supervisors keep a very messy office and claim it is a sign of being busy. Other supervisors keep a neat, well organized office and claim they are in control. How do you keep your office compared to your supervisor? If you are neat when she is messy or vice versa then she may not be able to relate to you. It is a good idea to mimic your supervisor in dress and office appearance unless your supervisor is clearly a maverick in the company.

How do You Present Information to Your Supervisor

There are generally two styles of presenting information. One style is the verbal way. For this method, people tend not to write instructions or reports but communicate the results verbally. The second style is the graphic way. For this style you present results in written form or transmit ideas by drawing pictures. A good way to find out what style a person prefers is by noting how they give directions. If someone gives you directions to a place by writing words (turn right, go north, etc.), he or she is usually a verbal person. If a person draws you a map, then he or she is a picture or graphics person. What is your supervisor's style? What is yours?

If you are a verbal person trying to report results to your supervisor and he is a picture person, he may have a tough time relating to you. He may be constantly asking you for something graphic to look at in your report. On the other hand, if you are a picture person and she is a verbal person you will have a different problem. She will not be interested in reading reports or looking at graphs. She just wants you to tell her the results.

Are You a Big Picture or Little Picture Person?

Some people like to talk details and not worry about the big picture. Other people like to discuss only the big picture. What type of person is your supervisor and what type are you? If you are both the same you will have more in common. If you have different styles you may have to consider changing yours.

Are You People Oriented or Results Oriented?

Some supervisors are very people oriented, which means they are really interested in you, your family, and how you are doing. When you interface with these people they want to hear about you just as much as they want to hear about work. So you want to remember to talk about yourself as well. Other supervisors are very results oriented and only care about work. They want to hear about what work you got done and they don't want to hear about anything else. How does your style relate to your supervisor's? If they are the same then you and your supervisor should be able to communicate openly and easily. If they are opposite, then you may want to communicate more in the style of your supervisor.

Are You Computer Literate or Illiterate?

How does your supervisor use a computer? Is he up to speed on the latest programs or does the computer sit in his office for weeks without being turned on? How does your computer style compare to your supervisor's; do you have the same style or are they different?

Fire Putter-Outer or Well Planned Out?

How does your supervisor handle business? One style is what is referred to as a fire putter-outer. In this style, she waits to the last possible minute and often works late into the night to meet impossible deadlines. This type of boss will expect the same from you—instant response and last minute panics to be taken care of. If you do not work well under these circumstances, you will have a hard time impressing your supervisor.

Another style is to plan everything out well in advance and avoid last minute panics at all costs. This style of manager must observe well planned and executed work—no last minute panics. A fire putter-outer working for this type of supervisor will always appear to be totally out of control and definitely not promotable. How does your style compare to your supervisor's?

These are some of the more common informal criteria that supervisors use to judge people. You need to be aware of these informal criteria and constantly checking for others that your supervisor may have. The informal criteria are often not obvious and they may not be easy to recognize, but they are there. With a little bit of research and acute observation you will begin to recognize other informal criteria. Your task is to start looking for and recognizing these informal criteria and use them to your advantage. In other words, take off the remaining part of the blindfold and see what you have to do to get the promotion.

ASSIGNMENT

1. Study the forms your company uses for job appraisals or job reviews. Do you understand each block of information and what it means in terms of your performance?

2. Who signs off on your job appraisal? Do they know you?

3. Make a list of all the things you were criticized for during your last job appraisal and identify something specific you can do for each one to show improvement. Then do it!

4. Meet with your boss and discuss the criteria you are expected to meet for your present job level as well as the performance expected from you for the next level. (Remember the cheat sheet.)

5. What is involved in the promotion review process for your company. Are totems used? If so when do they occur and what is your rank? Is a promotion review board used? Who are the members? What criteria are they using?

6. What informal criteria do you think your supervisor has? Can you recognize what styles are used? How do your styles compare? Are there any differences?

CHAPTER 7

HOW TO FIND AND DEVELOP A GOOD MENTOR

The dictionary defines "mentor" as a person who is a wise and trusted counselor or teacher. At all stages in our lives each of us depends upon and uses mentors. These mentors are usually referred to in different terms. When we are young our mentors are most often our mothers and fathers. As we mature other mentors come into our lives. These mentors may be teachers, coaches, or guidance counselors. Mentors may even be older brothers or sisters, or aunts or uncles. As we continue to mature our mentors might be called monsignor, rabbi, or professor as well as other titles. The point here is that every one of us calls upon and asks for the advice and guidance of mentors in our personal lives. Are engineering mentors necessary for successful career development? Yes! You need mentors at work just as you do in other areas of your life.

UNDERSTANDING THE BENEFITS OF GOOD MENTORS

It has been my experience and the observation of many co-workers that by finding and developing the right mentor you can significantly increase the chances for career success. If your mentors are well placed and well thought of in the organization, they can be your guide to the top, provided, of course, that you can meet their expectations and perform on the job. Mentors can help you develop the necessary technical and business skills that you need to progressively rise to the top.

A good mentor who is strategically placed in the organization can introduce you to the inner circle of executives who make the decisions as to who will be moving up.

Correspondingly, a mentor who is not strategically placed or not very well thought of in the organization, or offers poor guidance, can severely hurt your career. Your challenge is to find the right mentor and develop an amiable relationship. This relationship will significantly enhance your career development. In this chapter we will first identify all the benefits a good mentor can provide you, and then we will suggest ways you can find and develop good mentors.

A good mentor will help sponsor you, provide coaching, protect you, recommend you, and even help you get the more challenging, and consequently more rewarding, job assignments. Through these helpful mentoring steps, you will be exposed to the shortcuts and hopefully avoid the pitfalls and setbacks.

Sponsorship from a mentor is essential to getting the challenging assignments that provide you the opportunity to perform and clearly demonstrate that you are ready to move up. A good mentor is usually in a position with management to defend your abilities should they come into question for some reason. He or she is the person who interacts with management to vouch for your suitability to handle the difficult assignment. Your mentor is there recommending you for the assignment, clearly identifying you as the best candidate to successfully complete the assignment. It is this type of sponsorship that will make you stand out from the crowd.

Coaching is an equally important function that a mentor will perform. A good mentor can provide the coaching or polishing that is so necessary for advancement. Good coaching can help overcome a multitude of problems which you might otherwise have been unable to handle. Good coaching may come in the form of learning how to deal with an overly domineering supervisor or co-worker. It may also come in the form of technical advice as to how to solve problems or how to avoid technical failures.

Good coaching may come in the form of challenging assignments that cause you to expand your expertise. Or it may come in the form of recovery from failure, getting advice on what to do if you fail. A good mentor is a good coach, always there challenging you, inspiring you, and demanding the best.

Protecting is another function that a good mentor performs for you.

FIND AND DEVELOP A GOOD MENTOR FOR CAREER ADVANCEMENT

- **Mentor: Wise trusted counselor**

- **Benefits of a good mentor**

  - **Sponsor, Coach, and Challenge**
  - **Protection**

- **Finding a mentor**

  - **Older senior person**
  - **Outside your department**
  - **Feel comfortable sharing ideas**
  - **Similar styles**
  - **Advise having more than one mentor**

- **When to utilize a Mentor**

  - **Troubled times**
  - **New ideas needed**
  - **At beginning/end of a project**
  - **Recovering from failure**

Often people will try to find a cause or a person to blame when the engineering project fails. A good mentor will stand up for you in times of need and defend your actions, provided of course that he or she agreed with them in the first place. Mentors can easily deflect arrows or blame away from you. They can be a shield in troubled times.

By developing a good relationship with a powerful mentor you assimilate power just by being associated with him or her. You send signals to other people in the organization that you are a member of a powerful team. You acquire some of your mentor's influence and will have resources behind you. It may be possible for you to obtain inside information or new organizational power to cut through the bureaucracy and red tape.

FINDING GOOD MENTORS YOU CAN RELY UPON

The best place to start looking for a mentor is outside your department. Any senior superior in a position to influence your career positively is a potential candidate. Choosing someone inside your department may be quick and easy but will cause too many problems. Having a mentor inside your group can cause jealousy from other group members, innuendoes of favoritism, and even the alienation of other group members.

Finding a good mentor is not an easy task and may take quite a long time to cultivate [6]. When looking for a mentor you should be looking for someone with whom you are comfortable and compatible in ideology. The mentor should share similar views to yours about company strategies and success in the corporate world. The mentor should appear to you as a successful role model whom you would like to emulate. You should find the mentor stimulating, challenging, and an inspiration to you to perform to the top of your ability. This type of relationship cannot be planned but must be cultivated over a period of time.

The mentor should be a person with whom you can share your triumphs as well as your defeats. You will need him or her for guidance, non-judgmental listening, and constructive criticism. She or he should be a person with whom you feel comfortable trying out your new ideas. You should respect and value your mentor's honest opinion. He or she should also respect your intelligence and capabilities.

Unfortunately, mentors do not walk around the company with badges identifying themselves. So you must be aware of the subtle hints potential mentors give. When you are working with senior people do they take the time to explain to you everything you should know? Do they spend extra time making sure you get the assignment right? Do you find yourself sharing similar strategies on the best approach to the problem? Is there an unexplainable chemistry between you and your mentor? Do you enjoy discussing difficult problems and tasks with him or her? Does your mentor like your style and compliment you on your work? Similar outside interests are other things you might have in common with mentors. People who like your work and spread the word are good candidates. These are all good signs that the person you are working with would make a good mentor for you.

Once you have found a good candidate to be your mentor, the challenge becomes one of developing your relationship. Just as in any other relationship, you must invest your time. A good way to do this is by sharing lunch hours. Stop by your mentor's office on a periodic basis just to get an opinion about the project you are working on. Pass by your mentor's office on the way home at night. Volunteer your help on a project he or she is working on. If your mentor is like most people, he or she could always use a little extra help. Look for special projects around the company that are pet projects of your mentor's and get involved in them.

As you start to develop the relationship you can cultivate it by letting your mentor know how much you value his or her opinion. Everyone likes to hear how valuable their opinion is. You might even ask if he or she would mind if you considered them as a mentor. The point here is that you must be aggressive about finding a mentor and maintaining a relationship. However, you need to be cautious when you do this.

First, being overly aggressive can be misinterpreted as being too pushy or give the impression that your foremost interest is empire building. This can trigger resentment from your co-workers and earn you the reputation of being a "brown nose" or "apple polisher." Second, the relationship with a mentor must be beneficial to both you and him. You do not want to become just a "gofer" or "yes man." If your mentor is only using you for his or her own benefit, it's time to move on and look for another one.

Age difference is an important factor in picking a mentor. If your mentor is within four to eight years of your age, he is more of a close friend than a mentor. The best age difference is between eight and fifteen years of age. More than fifteen years of age, the relationship may turn into a parent–child relationship.

Don't put all your eggs in one basket. In other words, don't rely on only one mentor. If you spend all your time developing a relationship with one mentor and he or she leaves the company, you are suddenly left stranded. You will need several mentors. Even though mentors may appear to know it all, one mentor simply cannot provide all the guidance you need. For engineers it is good to have at least one technical mentor, one political mentor, and a mentor with great business acumen. This way you gain experience in dealing with technical, political, and business moves within the company.

For example, you may be working on a tough technical problem and require assistance from a senior technical person. In this case your technical mentor may be able to provide the best guidance. On the other hand, a project you are working on may be a dead end project as far as company management is concerned. A smart business mentor can forewarn you of the futility of working on the project and even assist you in moving on to more rewarding projects.

HOW TO UTILIZE A MENTOR

You must be aware that utilizing mentors can be overdone and underdone. You have to strike a balance, which will come with time and experience. If you are constantly seeking out the advice of your mentor for every conceivable problem, then you are overusing him or her. Your mentor will soon come to realize that you are incapable of making decisions and view you only as coming to him so he will do your work for you. On the other hand, if you only go to your mentor after you have solved everything and there is no real need to take advantage of his or her wisdom and advice, you lose the benefit of coaching. In either case your relationship with the mentor will not work. Here are some guidelines for when to utilize your mentor.

Times of trouble are probably the most obvious. When you are having difficulty and do not know where to turn, see your mentor. However, you should not "dump" your problems on your mentor. Coming to the discussion with some possible solutions indicates that your interest is in give and take to find answers to your problems. Be ready to discuss your options and highlight the pros and cons.

Ask your mentor for advice and find out what he thinks your options are. If you have missed anything helpful, your mentor will point this out for you. Ask for his or her guidance and be ready to act on it afterwards. Nothing is more discouraging for a mentor than to recommend a course of action and have the employee fail to even try it out.

Reporting back to your mentor on your progress is also important. Provide feedback on your activities; this shows that you are really taking and utilizing his or her advice.

Utilize your mentor for trying out new ideas and approaches to problems. A good mentor will have years of experience and should be able to assess what your chances are for success. Nothing helps sell a new idea faster than when your mentor is pushing for it along with you. In expressing new ideas, let him or her make constructive criticism, then implement any suggested changes. You may not see the need for the changes, but he or she may know of hidden barriers in the organization of which you are unaware. Your mentor's suggestions should help you to overcome these barriers.

A good time to utilize your mentors is at the beginning of the project. Meet with them and discuss your plans for the project, i.e., how you have set things up, your planned actions, and any problems you anticipate. Your mentor has years of experience and should be able to identify in advance your problem areas.

Approximately halfway through the project, discuss the problems you are encountering and the steps you are taking to solve them. Try to present problems with potential solutions. Ask for an opinion on how the project is coming along. Ask him to do research on how other superiors perceive the progress of the project. It's always good to know if management thinks you are doing a good job or an unsatisfactory one. Remember, mentors are great resources for finding out how to overcome barriers in the organization.

Another good time to seek out the guidance of your mentor is when the project is coming to a conclusion. What is the best way to end the project and present the results? How can you make management aware of the fine job you did on the project? What steps can you take to determine the next project you will be working on? Does your mentor have any recommendations? Can he or she sponsor you on another project? Who are the people that you must contact? These are all important questions that a good mentor should be able to help you with at the end of the project.

Another good time to utilize your mentor is when you feel that your position seems to be stagnating or when you feel pigeonholed. You feel that you are at a dead end in your job or have come up against some invisible barrier that seems to be holding you back. Good mentors will be able to tell you of things of which you may not be aware that are going on behind the scenes. It is quite easy for them to sit down with your supervisor or other people in the organization and get any information to which you may not have access. A good mentor will show you how to get out of the situation or just weather the storm.

When you have failed on a project and need to recover, never be afraid to ask for help. Some people react to failure by trying to hide it and project a superman image. Failure does not have to be a career limitation. In fact, a good mentor can show you how to overcome failure and actually make it an opportunity for advancement. Nothing is more impressive to management than when you can show them how you sought out the help of others, identified a solution, and made corrections after a failure. Doing this is not career limiting but part of career advancement.

Developing and utilizing mentors is a skill that you need to develop for career advancement. Knowing the best time to go to your mentor will depend upon your relationship with him or her. The above mentioned times are suggested as guidelines for helping you.

A WORD OF CAUTION ABOUT FINDING MENTORS FOR WOMEN

The world of engineering appears to be dominated by men. The chances are greater that most women will have men for mentors than other women. Unfortunately, good mentoring relationships between a male mentor and female subordinates are extremely hard to develop and maintain. Sexual interests, differences between male and female interests, and even jealous spouses or significant others may make your relationship difficult.

If you are a woman and develop a male mentor, be alert for the relationship shifting to something more than friendship. If this happens, then it is obviously time to change mentors. If you become emotionally or sexually involved it diverts your energies from your primary goal of career advancement. It can quickly ruin your mentor's career and yours. This type of involvement is to be avoided.

A good male mentor should challenge you and encourage you to make bold strides rather than timid little steps. His advice should be the same regardless of your gender.

Women looking for good mentors may want to take a different approach. Organizations such as Women in Engineering or Women's Engineering Societies are made up solely of women in engineering. Their goals and objectives are networking for women engineers, and they are therefore excellent places to find mentors.

Women mentors can provide tips on what it is like to be a woman in the engineering field. They can provide the guidance and coaching that is necessary for women to overcome the "Good Old Boys" barriers, how to handle sexism, and deal with the "overly friendly" male supervisor. For a woman in the male dominated engineering field, having several female mentors is an absolute necessity.

A FINAL NOTE

Most people become squeamish when faced with the task of developing a mentor. To this I say, what do you have to lose? The answer is absolutely nothing, and you have everything to gain. Remember, several good mentors are necessary for career development!

You will need to have many mentors throughout the course of your career. No one single person can provide all the guidance necessary. You may also have several different mentors at one time. As your career progresses your mentors will come and go just like they do in other parts of your life.

ASSIGNMENT

1. Make a list of several people in your company who you think might make good mentors for you. Pick one and approach him or her about

some problem you have and ask for guidance. Watch his or her reaction. Is it what you need in a mentor?

2. Can you identify a good technical mentor? How about a good business-oriented mentor?

3. What organizations outside of your company can you identify as good sources of mentors?

4. For Women engineers, contact your local Women in Engineering Society. Can you think of any other organizations that might be a good source of mentors?

5. Topics of Discussion:

Do mentors last forever?

How do mentors change as your career advances?

What are good qualities in a mentor?

CHAPTER 8

THE IMPORTANCE OF MAINTAINING A COMPANY CALENDAR

One of the most interesting phenomena that I have observed is the total lack of interest in the company calendar by most employees. Most engineers have a calendar on their desk or carry an appointment book to keep track of important events. However, if you ask them if they track significant company events or have a company calendar, the answer is almost inevitably negative.

People have calendars for school events, sporting events, and even for birthdays and anniversaries, but rarely have company calendars. Is it important to develop a company calendar and keep track of significant events? The answer is yes!

> Remember: failing to plan is simply planning to fail! And not having a company calendar is planning to fail.

WHAT GOOD IS A COMPANY CALENDAR AND HOW DO YOU USE IT?

To show you how important developing and maintaining a company calendar is, let's look at the following example. Shown in Figure 8-1 is an

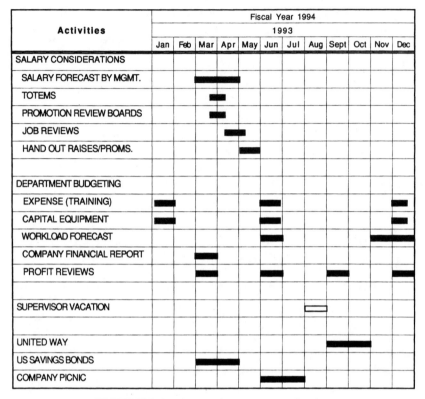

FIGURE 8-1   A sample company calendar.

example of a company calendar. Let's study the first significant event on the calendar: salary considerations. In this company, raises and promotions are given out once a year, in May. However, a significant amount of activity occurs before the May raises.

In the March through April time period the supervisor puts together a preliminary salary forecast. This is the supervisor's first estimate of the raises and promotions intended to be given out in May. Occurring also in this time frame are the company totems and promotion review board meetings. Toward the end of April, job reviews occur. Last is the actual awarding of raises and promotions. Now let's examine why keeping track of these events is so important.

Let's assume that you have been working for several years without a raise or promotion and you finally decide that its time that you do something about it. However, it is September and promotions or raises are only given out in May. What good do you think will come of you storming into your supervisor's office in September and stating, in no uncertain terms, that you deserve a raise now. An immediate response is to stall. Your supervisor knows there is no possibility that you will receive a raise at the present time but the worst thing to do is tell you that now. How much better it would be if you have saved your energy and waited for March.

The next thing to consider is job reviews. Your supervisor is busy handling a multitude of problems. Suddenly its April or May a job appraisal must be written for ten employees. If one appraisal writeup takes 3 hours, it may take over a week just to fill out the forms. Does your supervisor have time for this? No! He has ten employees and probably can only remember about 20% of what they accomplished during the past year. This time period is not the time to make waves or pick a fight with your supervisor. If you recently had a setback, chances are that this will be foremost in his mind. The best thing to do is have your breakthroughs and successes timed so they occur in March.

One way around supervisor forgetfulness or your recent setbacks is to send your supervisor a memo in March summarizing all your accomplishments over the year. Simply type up all your accomplishments and go over it with your supervisor. Remind him of your tremendous effort during the past year and highlight all that you accomplished. Inform him that you did this to save him time when filling out the job appraisal forms and he need only file it away until he is ready to do yours. It should help and save time. Chances are he will do just that. He will welcome you making his job easier. Make sure you keep a copy in case he should loose or misplace his! Even sending his supervisor a copy of the memo is a good idea.

Let's discuss some other important activities that are occurring at this time. Company totem is one and promotion review board is another. Do you know who is on the totem board? Do you know who is on the promotion review board? If you do it is a good time to beat your drum a little bit so they remember your accomplishments over the past year and not just the latest one, which occurred last week. Those rated highest on the totem are usually the ones who just recently made a major accomplishment. Do you know the criteria the promotion review board will be using? A ghost memo written by you for your supervisor that highlights your accomplishments against the criteria is very handy to have in advance.

> Timing is very important. Successful career people do not depend on luck. They make their own luck, and one is through the company calendar.

Another important activity on the company calendar is the distribution of internal research and development funding. Do you know when this occurs? For the example shown, initial planning is done in January and progress reviews are held quarterly throughout the year. Do you have a project you would like to get funded? What projects are funded for the year and who is working on them? Are you on a pet project of great importance or something to do to keep you busy? Knowing when the project reviews are planned is important information. Nothing helps like having a break-through on your project just in time for the Vice Presidential review.

When do United Way and the annual Saving Bonds Drive occur in your company? These are excellent projects to donate your time to. These activities are often organized and executed by upper level managers—the same managers who will be sitting on totem boards, promotion review boards, and handing out raises. This is an excellent way to get the extra visibility that you need. While working for a noble cause you have the opportunity to go around your lead engineer and supervisor and meet their supervisors on a one-to-one basis and get extra points for doing it. As you sit there planning out the activities with the upper level management it doesn't hurt to tell them about all the problems you are solving.

Other examples of activities like these are the company picnic, the company holiday party, and any charitable things your company may sponsor, like food drives, shelters, clothing drives, or marathons. If you keep a company calendar you will soon realize that there is something happening nearly every month that can result in giving you the extra visibility that you need. So much in fact, that you may not be able to take part in it all. It's important to make new contacts as well as friends along the way. Developing genuine friendships is as important as developing one's career. You can do both.

Another important activity you need to be aware of are the workload forecasts that your supervisor must do on a periodic basis. These are simply forecasts showing the work planned for you and your group over the next year. When does your supervisor do these? It may be a good idea if you paid attention to these, since they will tell you if you will have a job in six months. If your supervisor is forecasting enough work then, come job appraisal time you stand a better chance of getting the raise. If, however, in three months the forecast is no work for you, maybe it's a good time to start looking for work elsewhere. Remember the old saying, "God helps those who help themselves."

Another annual activity is updating the business plan or the company five year plan. This plan identifies the key business thrusts that the company plans to execute over the next one to five years. Knowing that such a plan exists, when it gets updated, and obtaining a copy can be very beneficial to your career.

Department or supervisor budget planning is another activity of which you should be aware. A supervisor must forecast or estimate the expenditures the department will have over the year. Periodic updates and adjustments are made throughout the year. Stopping by your supervisor's office when a budget is being done is a good idea. Was your tuition for classes you planned on taking included? Has the budget allowed for your new computer or the test equipment that you need?

Another important activity, which usually occurs on a weekly basis, is staff meetings. Do you know when your supervisor must meet with superiors and report progress. If you just happened to stop by and brief your supervisor hours before walking into a staff meeting, she may be more inclined to highlight your accomplishments. It does not hurt to have your supervisor highlighting all your accomplishments at the weekly staff meeting. Another way to do this is to submit a weekly written report to your supervisor highlighting your accomplishments for the week. Remember to save these weekly accomplishment lists. At the end of the year they become a handy summary of all your accomplishments. These weekly reports become a diary from which you can write your yearly summary.

If you don't write down everything weekly you will probably only remember about 50% of what you accomplished for the year. This situation gets worse when you fail to summarize your yearly accomplishments in a written document for your supervisor. Without something written down, your supervisor will only remember 30% of your verbal message when you next meet. This means that without disciplined and documented summaries of your accomplishments, your supervisor will at best only report 15% of what you accomplished. No one gets raises or promotions when the supervisor only remembers or documents 15% of your efforts.

The important point about developing and keeping a company calendar is that it allows you to keep track of important career activities, activities that you can influence so as to benefit your career or take actions necessary to maintain your career. Developing a company calendar is no easy task and it may take you several years before you are recording all significant activities. Once you do this you will soon realize all the opportunities that are available to you. It will quickly become apparent that the only person limiting your career growth is you!

ASSIGNMENT

1. Start a company calendar.
2. From the calendar, determine the most significant thing you can do in the next two months to enhance your career.
3. Volunteer for a meaningful or noble company activity. Get involved!

CHAPTER 9

THE LEADING REASONS WHY ENGINEERS FAIL

Most engineers were rated in the top quarter of their high school graduating class. Admission to all engineering schools requires above-average grades in math and science. In order to graduate, the average engineering student must spend four long years studying hard a variety of extremely complex and abstract subjects. Therefore, one would expect that these highly trained and intelligent engineers would not fail once they leave school.

However, this is not the case. Engineers do fail, and more often than is realized. Engineers fail on the job principally because they do not develop the broad outlook and basic human-relations skills that are so important to achieving in a team environment. In this chapter we will explore some of the most common reasons why engineers fail and how you can avoid making these mistakes.

DEFINITION OF FAILURE

The definition of failure takes on many different meanings for people. First, it is necessary to describe what failure is. If a person is reassigned or removed from a project against his will because he is too disruptive to the team or fails to get the work done, this is failure. A more drastic failure is termination from the company or to be "pink slipped," as it is sometimes called.

Failure may come in more subtle forms. For example, success is thought of as moving up the corporate ladder. Therefore, stagnation at the same level year after year rather than advancing can be considered failure.

Receiving minimal raises, being assigned trivial duties, constant reassignment, or continual transfer are all forms of failure. Being pigeonholed into one job or only one type of assignment can also be considered failure. There can be technical failure as well. The engineers fails to solve the problems and the company losses a significant amount of money. Or the engineer may fail to get technical credit for his work.

The definition of failure is different for every person. What is common about all the different types of failure is that the engineer does not reach the goals he or she intended to reach and there is no career growth. Let's explore some of the common causes of failure for engineers.

LEADING CAUSES OF FAILURE IN ENGINEERS

Studies have shown that there are a wide variety of root causes for the failure of engineers [7]. Probably the most common and noteworthy among these are:

1. Inept or poor communication skills
2. Poor relations with the supervisor
3. Inflexibility
4. Poor and lax work habits
5. Too much independence
6. Technical incompetence

These reasons for failures were highlighted since they cross all technical fields and are probably the most prevalent ones cited on job appraisals. They are the problems that are likely to continue to haunt you throughout your career unless you aggressively do something to correct them.

IMPROVING INEPT OR POOR COMMUNICATION SKILLS

Good communication skills are absolutely necessary to move up in the company. The inability to effectively communicate is what often keeps an

---

CAUSES OF FAILURE FOR ENGINEERS

1) **Inept or poor communication skills**

2) **Poor relations with the supervisor**

3) **Inflexibility**

4) **Poor and lax work habits**

5) **Too much independence**

6) **Technical incompetence**

---

engineer from advancing. Good communication skills are required in a variety of areas. For example, good communication skills are required for reporting progress to management, giving specific direction to subordinates on things to be done, dealing with customers, describing complex problems over the phone, presiding at meetings, writing specifications or reports, and even interfacing with people over computer networks and in video satellite conferences.

Engineers get paid to resolve complex problems using communication skills to bring together the resources, people, and technical knowledge necessary for success. Lack of good communication skills will obviously limit the success of the project and your career.

By nature and training, engineers tend to focus on technicalities rather than people, and this often tends to make engineers poor communicators. Many believe technical skills are all that count and that they will be rewarded accordingly. Technical skills do make up a portion of the criteria for advancement but communication skills are also another part of the criteria.

An engineer who is able to communicate clearly (in writing and verbally) and has a good technical understanding is usually recognized by superiors as someone with high potential. To illustrate how important good communication skills can be, let's look at some examples.

Clear, concise, and easily understandable writing is a must. Engineers must often write specifications and technical reports. Specifications usually define requirements for a product or process. Poorly written specifications for these products or processes can cause a multitude of problems. First, poorly written specifications need to be interpreted and rewritten so that the true meaning becomes apparent. This can cause delays in the start of the project, because it appears that the product is being designed to unsafe or potentially life-threatening specifications. All of this may result in cost and schedule overruns, potential lawsuits, and even death.

Poor technical writing skills can be improved upon, but it takes time, practice, and more practice. One way to improve your writing skills is by taking technical writing courses at your local college or university. Another thing you can do is look for guidance in old reports, specifications or other documents previously generated by other people in the organization. Adapt their outlines, forms and style of writing as much as you can. Getting your hands on an old report that was well received can save you hours of rewriting and editing.

Clear and concise verbal skills are a must. Career progress, in part, depends upon an ability to sell oneself and one's ideas. You need to be able to do this during your meetings with management as well as with fellow employees. You must be able to verbally communicate the important points contained in the technical charts, graphs, and reports that you have written. You must also be able to verbally give clear and concise instructions to people so they know what must be done. This is especially true when working on potentially dangerous projects such as nuclear reactors, high voltage equipment, or with corrosive and poisonous chemicals.

Developing good verbal skills takes time, but you can take advantage of shortcuts. Listen and study how persons in the upper echelon of the organization report verbally. On what do they put the most emphasis—cost, schedule, or technical aspects? What is the standard form for presenting verbal results? Do they use overhead projectors, handouts, drawings, computer plots, or some other type of graphics to support verbal reports? Seeking out opportunities to speak before groups or perhaps present technical papers is excellent training.

There exist organizations which help to improve speaking skills. Probably the most common of these is the Toast Masters. This is a national organization with local chapters dedicated to helping people learn how to make speeches and presentations in front of a group. By studying the speaking styles within your company and through practice you can significantly improve your verbal communication skills.

Poor verbal and written skills can be overcome, but this will take time. Once an engineer has made a poor impression with written reports or verbal presentations, she may not be asked to assist in certain key activities because of that impression. For this reason, communication skills are extremely important for the new engineer during the first year. Communication skills may even rank higher than technical skills, since new engineers are not given the more technically challenging tasks.

TURNING AROUND POOR RELATIONS WITH THE SUPERVISOR

The personal relationship an engineer has with his or her supervisor is probably the dominant factor in determining success or failure. Many engineers, especially recent graduates, do not understand this and greatly underestimate the importance of their supervisor. They naively believe that if they do a good job, their supervisor will always recognize it and they will be successful. Doing a good job is not enough. What counts is doing the *right* job and having your supervisor recognize this. Previously, it was highlighted that in the business world there may be many solutions to a problem. The challenge is to pick the right solution. This is where the engineer–supervisor relationship comes in.

In order to advance you must have a good relationship with your supervisor. Engineers must be able to discuss problems, report progress or lack of it, identify solutions, and, finally, get the supervisor's approval. The engineer must understand his or her role on the team and relationship with the supervisor as one of cooperation and providing assistance.

The quality of your work will greatly suffer if the relationship with your supervisor is not good. Your supervisor is not your enemy. He has been there before and has a pretty good understanding of what needs to be done and how to do it. He needs your cooperation and support, not a strained and troublesome relationship. This relationship with your supervisor should be similar to that of a hand and a glove. The hand and glove go together. You and your supervisor work tightly together and depend upon each other to get the work done.

How would you rate your relationship with your supervisor? Would you classify it as good? Would you classify it as bad? Or have you never really thought about it? These latter two are failure relationships. If things are going well between you and your supervisor you should be able to openly and candidly discuss work and problems. You should feel you can ask for guidance and it should be given willingly. Your supervisor should be defining a course of action and you should be implementing it. Not all relationships run that smoothly and there will be disagreement. The important thing is that you can air your differences and move on. Most times things should be running smoothly.

There are danger signs to watch for in your relationship with your supervisor. The first is that the two of you are usually in disagreement. Does he always seem to be saying white when you're saying black? Is he or she constantly nagging and deriding you and discounting your work? Does your supervisor continually redo all your work with no explanation? Does he continually give you a sense that you are incompetent? If you answer yes to any of these questions, then it's time to realize you have a very poor relationship with your supervisor, which can be career limiting for you. It's time to sit down, and air your differences.

Try to find a common ground. Identify what you can agree upon and concentrate on this rather than on how much you disagree. Identify what you can do to help the situation. Ask for inputs. Do not ignore the situation; it will only get worse.

If the poor relations are caused by different technical opinions, the best thing to do is to at least agree to disagree. This is where you both agree that there is an honest difference of opinion. You are each declaring that you respect the other's opinion but you disagree. By doing this you are allowing each other to save face. Neither of you is right or wrong.

If you disagree with your supervisor on your performance it is a different matter. To change an opinion of your performance you will have to start performing the way he wants you to and not the way you want to. You will have to clearly demonstrate actions that meet any criteria. Changing your supervisors opinion will be hard and it will take time. It will also take a large amount of effort on your part. If, after trying all this, you feel that you are not making any progress it is time to move on and look for a new supervisor.

OVERCOMING INFLEXIBILITY

There are several sayings that characterize inflexible behavior. The first is "There are three ways to do the job, the right way, the wrong way, and my way." The second is "If you want things done right, you have to do them yourself." And the third can be found on a common sign displayed around offices. The sign reads:

RULE 1   The Boss is always right

RULE 2   When the Boss is wrong see Rule 1

If you find yourself agreeing with these approaches to problems, chances are you are very inflexible. It may even be a problem that could be limiting your career advancement.

All these point to inflexibility, which is another cause of failure by engineers. Inflexibility is subconsciously taught in colleges and universities. Engineering students are taught that there is only one correct way to work through a problem and derive the correct solution. The object of most classes is to learn how to apply well-defined rules and formulas to come up with the one right answer. All other methods lose points and your grade drops. On the job, however, problems are not well defined and there exist a multitude of correct answers.

Solving problems on the job requires a team effort, with inputs and solutions suggested or derived from many people. Often cost and schedule do not allow you to score a perfect 100. In fact, the final solution may be far from optimal. All this requires the engineer to maintain a balance in his work. Remain as flexible as possible and with time and experience you will learn to find the optimum solution.

Being inflexible can cause the team and you a multitude of problems. Most people will pull away from someone who is too inflexible. If you are inflexible others will feel that you will only implement your own solution to the problem and their suggestions do not really count. This is failure for a team leader. Clearly, working on a team requires that you consider inputs from everyone and choose the best solution regardless of who's idea it may be.

Engineers have different backgrounds, styles, and formal schooling. Therefore, you must be able to work with each of these different styles. Flexibility is the key word. You must change your style in relation to those around you to stimulate them as well as yourself to get the best performance. An example that I like to use is that of the coach of a successful football team who displays a different style when talking to the defensive line than he does when talking to the quarterback. He must be flexible and change his style in order to get the most from his players. Similarly you must be flexible to get the most for your career.

There are several questions that you can ask yourself to help determine if you are flexible or inflexible. If you are presented new information, do you take the time to evaluate it or simply discount it since you already have the best solution? Do you tend to deal with all people and problems in the same way or do you try to tailor your approach to that particular problem or circumstance to attain the best result? Do you ever change your mind?

Do you strictly follow every company policy, procedure, and regulation with no exceptions? Is there and has there always been only one correct way to get the task accomplished?

Flexibility is key to career growth. Inflexibility is only good when it comes to compromising personal moral or ethical standards.

CHANGING POOR OR LAX WORK HABITS

Poor or lax work habits as well as work schedules are other reasons why engineers fail. Poor or lax work habits and schedules can take many different avenues. Are you easily distracted as you go about your work day? Do people come and go continually in your office making you feel that you just never seem to have the time to complete the work at hand? Are you unable to concentrate for very long on a problem before you find yourself wandering off to another problem? In other words, do you jump to the next problem or task before you have completed the present one? Do you get caught up in all the details and never seem to find a solution? Does it seem that there is always some little detail coming up at the last minute to nullify your work? These are all signs of poor and lax work habits. Habits which, if left unchecked, will cause career failure.

The reason these habits are so career limiting is the dramatic effect they have on the quality and quantity of work you accomplish. Being easily distracted can turn an easy, short task into a major one. This type of performance causes cost overruns and missed schedules. Jumping from problem to problem before you complete the last one leaves you with nothing ever being accomplished. You have worked on everything as the supervisor wanted you to, but you have completed nothing. Again, a failure mode for the engineer.

Getting caught up in the details and wandering aimlessly is very career limiting. The engineers of today are in the middle of an information explosion. They must be able to sort through massive amounts of information and determine what is important and what is not. In order to do this they must be well disciplined and organized. This means having methods of sorting and storing information, keeping good notebooks, and summarizing whenever possible. It means that you have the ability to organize and run complex experiments and keep track of all variables.

Lack of discipline and being disorganized are two major reasons why most projects fail or end in cost overruns. I cannot believe the number of times experiments or tests had to be rerun simply because the engineer was not organized or failed to pay attention to details or never bothered to record test conditions or results. Poor and lax work habits cost the company time and profits. They can even lead to injury or death in extreme cases.

AVOIDING TOO MUCH INDEPENDENCE

Most engineers are members of a design team. Team members must work closely with each other and draw from each other in order to accomplish their goals. If an engineer becomes too independent the team can suffer. Some engineers believe they can do it all and try to take on bigger assignments than they are capable of handling. Others want to work alone with no one overseeing their work. In either case these engineers often go off into a corner of the company and get stuck because they are too timid or have too much ego to not ask other team members for help. The net result is that their failure to solve the problems is often discovered too late by the team. Everyone is affected and the outcome is failure.

Don't be too independent! It is too career limiting. Try as best you can to be a team player. Take the time to share results with your other team members. Do not be afraid to ask questions and get clarification if you need it. You may not have all the solutions but neither does anyone else on the team. By sharing ideas and discussing problems you may find that someone else may have the answer you are looking for. Better yet, you may have the answer to someone else's problem. You will find the solutions through sharing and working together as a team. Remember, any team, regardless of the sport (football, baseball, volleyball), needs all of its team members working together to score. Your work team is no different.

If you are a loner it will take some time to make the transition to team efforts. Take things slowly at first and, at a minimum, at least show up for team meetings. If you are afraid to share your work in front of everyone, then schedule special time to review your work with only the team leader. Try to find at least one other person on the team with whom you feel comfortable sharing results. The point is that you must change to survive, and only you can do it.

If you are a team person but are experiencing problems working on the team, try to discover why. Is it personalities, different backgrounds, or the way the team is organized? Try to identify specific things that are causing you problems. Ask if others are having similar problems. You may find that you are not alone. Some teams gel and everything clicks. Other teams develop a multitude of problems and try your patience to the limit. Stick with it, teams do not last forever. There will be another team and other tasks. The best way to move up is to do your best on the present task. And prove you are ready for a more challenging one. If you're not up to the challenge of the present task, what makes you think you will do any better on a more difficult one.

GETTING HELP AND AVOIDING TECHNICAL INCOMPETENCE

Last on the list of reasons why engineers fail is technical incompetence [8]. This comes as a surprise to most people. An engineer must learn to face technical challenges to accomplish two things. The first is to broaden his technical knowledge and experience base. The second is to technically update himself periodically throughout his career. To neglect either results in failure in the long run.

The products of today are extremely complex, requiring a multitude of engineering disciplines working together to successfully get them out the door. Take, for example, a copy machine. This machine requires that electrical, mechanical, optical, and chemical processes must all work together. If you expect some day to advance to the position of team leader, you had better expand or broaden your background to be able to deal with each of these disciplines. To be a really good team leader you must understand the customer. This may mean looking into sales and marketing and broadening your knowledge of these fields also. The point here is that to continue career development you must broaden the scope of your knowledge.

If you prefer to remain competent in one area only, then you must periodically update your knowledge in that field. This updating may take the form of attending seminars or returning for classes at a university. Why is this necessary? Because the world of engineering is rapidly changing. Not updating yourself results in failure in the long run.

SUMMARY

Being successful in college does not guarantee success in the business world of engineering. Successful engineers can fail once they leave college. The most common causes of engineers failing are poor communication skills, poor relations with the supervisor, inflexibility, poor work habits, too much independence, and technical incompetence.

Failure can take many forms. Some of these forms are: stagnation at one level in the company, minimal raises, constant reassignment, poor assignments, and, ultimately, being fired. Failure is never final, and with effort you can turn your performance around. To do so you must work at it on a daily basis, constantly improving yourself.

CHAPTER 10

GETTING PEOPLE TO
ACCEPT YOUR IDEAS

Most engineers will probably generate several hundred ideas during their career. Some of these ideas will be wonderful solutions that could earn an award or promotion. Unfortunately just coming up with the idea is not good enough. No matter how good the idea is, it will not sell itself. You must be able to convince other people of its merit otherwise it will not benefit anyone or your career.

Engineers must have the skills to sell ideas to other people [9]. Convincing other people of the merits of your ideas is not a natural talent but a learnable skill. This chapter will show you how you can identify career advancing ideas and develop the skills you will need to sell your ideas to other people. Your ultimate objective should be to learn these skills to benefit your company, coworkers, and your career.

GENERATING CAREER ADVANCING IDEAS

The first step is coming up with a new idea. This is not an easy process and there are no guaranteed rules you can follow to generate great ideas. You can aid the process by first being alert or sensitive to problems in your daily work. If you perform your work every day with the attitude that what has worked in the past will work today, you will not discover very many career advancing ideas.

You must think continuously of improvement—continuous self improvement and continuous job improvement. Every person and every job can be improved. Adapting this attitude is the start of generating career advancing ideas. Do not be satisfied with the status quo. Always be on the lookout for new and better methods [10,11].

Being aware of how you can improve yourself. Your present job is your best place to start. Think of the daily tasks you perform and the problems associated with getting your work done. Can anything be improved upon? Chances are there are going to be several. Start with simple things that are directly within your control and which you have the power to change. If you can implement a new procedure or create a new tool to save time, do it. Then show your supervisor how you improved the situation. The key is to start small with your present job and then move on to bigger and better ideas, all the while gaining experience and knowledge on how to sell your ideas. You must start with improving yourself and then move outward to other areas in the company.

Many junior engineers will immediately start questioning everything that is done in the company and quickly come up with better ways to improve everyone else around them. This is not what you want to do. I have on many occasions had to point out reality to junior engineers about their great ideas to change the company. I tell them point blank, "Demonstrate to me that you can first do your present job better and quicker; after that I'll listen to your ideas on how to improve other things!" "If you cannot even improve your present job how do you ever expect me to believe in your other ideas?"

Once you have demonstrated that you can come up with good ideas to improve your present job, it's time to expand your horizons. The next area to look for ideas is when you are working on interdepartmental problems. Look for new ways that may help two different departments to more easily accomplish their work. A simplistic example is the combining of two forms into one.

Often different departments will have you fill out forms to document the type of work you are requesting from them. Usually every department has its own customized form. It would save time to have one standardized form that both departments could use. It would save time by having to only fill out one form and decrease the company's printing costs by reducing the number of forms from two to one.

Another simplistic example might be changes to your design that allow the product to be assembled more easily. You might tour the production line to see how your particular component of the product gets assembled. If you could identify a simple change that would save assembly time and not sacrifice quality, this would be a great improvement.

Another good source of ideas is your supervisor, and there is no better place to start looking! Supervisors have pet ideas that they would like to try out to see if they would work. However, since they are too busy just running the office they often do not have time to try anything. Why not volunteer to help your supervisor's ideas? It is a well known fact that ideas "friendly" to management are the ones most likely to be accepted and implemented.

There are several ways that you can benefit from suggesting ideas. One method is through the company sponsored cost cutting program. Nearly every company has such a program. These programs are set up by management to help encourage the submission of cost cutting ideas.

Companies usually have a form you fill out and submit to a committee for review. Get some of these forms and become familiar with how to fill them out. Learn the criteria of the committee that judges the ideas for awards. Keep the forms handy; put one up in front of your desk so you can look at it every day. Try to submit at least one or two suggestion per year. Make this your goal.

"Why bother," is the response I most often hear, "they will never accept my idea!" To this I reply, "What do you have to lose?" The answer is nothing, you only have something to gain. The truth of the matter is that you have more of a chance of winning an award from your company than you do winning a lottery. I know of people who will spend hours of effort trying to come up with the magic numbers for megabuck lotteries they have very little chance of winning. If they put that same effort into submitting cost cutting ideas, they would have significantly better chances for winning at work.

Is the effort worth it? That depends on how much the ideas can save money. Some companies offer as much as 10% of the cost saving to the suggestor as a cash award. On an idea that saves $10,000, this could amount to $1,000. To me there is very little effort to writing up an idea, less effort than driving to a store and purchasing a lottery ticket. In fact, it does not cost anything to submit your idea, and you have a much better chance of winning. In addition, most companies pay the tax on the award so you are not penalized; with the lotteries the IRS is there the minute you collect your prize. Hopefully this will convince you to get moving.

Another benefit of suggesting new ideas is career advancement. Often engineers are in position to create new products for the company or significantly improve the quality of an existing product. These ideas cannot be directly rewarded with a cash bonus, since it may take several years to implement them. The company therefore rewards these engineers through career advancement. For these types of ideas, the engineer cannot simply fill out a form; they require a different approach. Let's now discuss a different approach you must take if you plan on getting people to accept your ideas.

THE IMPORTANCE OF PUTTING IT IN WRITING

If you come up with a major new idea, the first thing you need to do is get organized. And the best way to get organized is by writing your idea down. The worst mistake you can make in trying to sell your idea is to rely upon verbal communications. The people you are trying to sell your idea to, particularly your supervisor, are bombarded with new ideas all the time. They are not going to take notice of what you have to say unless you give them a good reason. The way to do this is by putting your idea into writing.

By putting it in writing you are sending an important message to everyone. The message is that you think the idea is so good that you have put special effort into writing it up and documenting every detail so you don't lose it. Remember, your supervisor will only hear about 60% of your idea if you tell it verbally. After a few days he or she will probably only remember about 20% of what you suggested. And after a month, you can bet it's been totally forgotten.

Your supervisor is like everyone else—far more interested in what they have to say than in what you have to say. A verbal exchange of ideas will probably leave you frustrated because of your supervisor's inability to grasp the magnitude of your idea. If you think your idea is worthwhile, then give it the emphasis it deserves. Write it up! There are several benefits to a neat, well organized write-up.

1. By writing it, you are going on record. You are showing everyone you think this is important. Written suggestions immediately invoke a need in the supervisor to seriously consider the idea.
2. It causes you to get organized and think your idea through. Remember only you have a crystal clear vision of your idea; everyone else must rely on your verbal description, written words, or pictures.

When you write something up it causes you to put your idea into the simplest terms—the best way to communicate it.

3. A written description serves as a reference for future discussions. It is also the first step in generating patents. Light bulbs go out as fast as they come on. Without a written description it is easy to forget all the reasons why your ideas are so deserving of consideration.

4. Writing a description tends to eliminate the personal and emotional overtones that often accompany an engineer's idea. The emotional highs and excitement you have about your idea can turn people off before they have had the chance to really evaluate its true merits.

When preparing a write-up or proposal, your objective is to prove that you thoroughly understand the problem and have a great solution. When you write up the presentation you must establish enough credibility to show you have seriously analyzed the problem and have a solution that represents a significant improvement. Only then will your supervisor have enough confidence in you to really pay attention. To establish this type of credibility in a write-up or presentation you must first do your homework.

WHY RESEARCHING YOUR IDEA IS NECESSARY

In order for your written description or proposal to have any credibility, you must support it with research. Your research should answer the following seven questions.

1. What is the problem and how big is it? An easy mistake is to assume the problem is apparent and your supervisor perceives the need to solve it as you do. First you must make sure he is aware of the problem. To do this, describe the problem in simple terms. Get his agreement that a problem exists. If you are intent on solving a problem that he considers trivial, no matter how good your idea, he will perceive your solution as trivial.

   Next, explain the circumstances that exist that make the problem worth solving. Quantify how big the problem is in terms of man-hours lost, dollars lost, or poor performance. Supervisors take immediate notice when the problems are in terms of cost overruns, failures, schedule delays, or rework required. Once, you have clearly identified the problem and assessed its impact it's on to the next question.

2. What if nothing is done now? Quantify the short term and long term impact of the problem. Most managers will hope the problem will go away if ignored. Show the manager the consequences of ignoring it. Identify any new problems that will arise if the present one is not solved. Once the magnitude of the problem starts to settle in, he or she will want to know why your idea merits consideration. In order to answer this you must do some further research. Your objective is to establish credibility in your answer, and to do this you should have knowledge of the past history of the problem. This leads to the next question that is normally asked.

3. How did we get here? To answer this question you must do research on what was done up to this point that lead to the problem in the first place. Map out or list the decisions made previously that lead to the problem. You may discover that the problem you are trying to solve is small in comparison to other problems.

4. What was suggested before? Try to discover any past ideas that may have been suggested to solve the problem. Your objective is to show that your idea does not take them down a path that has previously been tried and failed.

5. What has not been tried? Here you want to summarize several ideas that show potential for solving the problem, your idea included. Your aim is to show that your suggestion is the best. If you really do not have a good solution, don't fake it. Drop the idea. Overselling the benefits of your idea and not thoroughly thinking them out will destroy your credibility.

6. What puts your idea ahead of the rest? Highlight your idea by researching what is great or different about it. Summarize in simple terms the benefits of your idea. Quantify all the benefits you can. Supervisors like to hear the benefits in terms of cost reduction, lower unit-product cost, improved performance, quicker turnaround, improved producibility, product reliability, and product safety. These are few of the buzzwords used to help sell ideas.

   After you have completed all the research on the benefits of your idea, you need to also consider the bad points. A well thought out plan will not only present the good points but also the bad points. All new ideas have problems associated with them. Everyone knows that nothing is free and everything costs something. Your supervisor is no different, and will look for potential problems or snags.

   The best way to deal with this is meet it head on. Do no try to hide it. Make it part of your plan. Identify the problems with your idea.

If possible, show that the consequences are trivial and the benefits outweigh the negative effects.

7. What will it take to implement your idea? If you have a great idea, the first question your supervisor will ask is "what will it take to implement it?" This is where you can really establish creditability by having done your research. Do research and make estimates to show that your idea will not break the company or the budget. Explain how funds can be redistributed to handle what you propose or how long the payback will take. Is what you proposed feasible in terms of schedule? Are the resources and manpower available? How will it affect the group's morale and customer relationships? Are any special facilities needed? Identify the expertise and skills needed as well as where they can be found.

People often come up with excellent ideas that are not within the charter of the group or company. An advantage to selling your idea is to show how the idea complements the company strategic plan or present product lines.

With your research completed, it is now time to put together the presentation or plan.

WRITING UP THE PLAN

The exact organization or style of the plan you put together will depend upon your company's products and organization, your supervisor's style, and your own preferences or style [12]. There is no right or wrong style. The length is entirely up to you. If you are proposing a multimillion dollar plan with a hugh impact to the company, you may want a detailed and thorough plan. These plans may consist of 30 to 40 pages. On the other hand, if you are just trying to make a simple change in your department your plan may be only a couple of pages long. To aid you in writing up the plan I have put together the following outline:

1. Executive Summary
2. Statement of the Problem
3. Background/Previous History
4. Potential Solutions (Pros and Cons)
5. The Best Solution
6. Benefits and Impact of Implementing

7. Implementation Plan

8. Conclusion and Summary

Executive Summary. This is a summary of all the high points of the plan. It clearly defines all the benefits of the plan right up front. Your intent is to get your supervisor so interested in the idea or plan by the executive summary that he or she has to continue reading it to find out more.

Statement of the Problem. In the simplest terms and shortest possible means, describe the problem as best you can. Draw diagrams if needed, show charts or graphs that pictorially describe the problem. Use the material and information that you have collected during your research to define the problem and show how large it might be.

Background/Previous History. Next present a summary of the background leading to the problem and actions taken previously to solve it.

Potential Solutions. In this section list or identify all the potential solutions to the problem including the one you are recommending. Explain what is beneficial about each potential solution.

The Best Solution. This section should contain a concise, detailed description of your proposed solution. This is where you expand upon your idea. Discuss the details and highlight all the benefits.

Benefits and Impact of Implementing. This is where you summarize all the research you have done on the cost and impact of implementing your idea. Organize it in a neat, concise manner to address all the issues that may come up.

Implementation Plan. In this section you clearly identify the plan that must be executed to implement your idea. Here is where you present the research you did, and what it will take to get the proposed solution or new idea working. The best method for doing this is through a schedule showing the tasks to be completed, the sequence and time phasing of the tasks, and the manpower and costs associated with each task.

Summary. This section should be a one or two page summary, at most, that describes all the highlights of the previous sections. It should be a section of the write-up that your supervisor can go to and get an executive summary of the points of your plan. It should clearly state that you recommend implementing the proposed idea.

HOW TO BEST PRESENT THE PLAN OR IDEA

Once you have completed the plan your first thought will be to rush into your supervisor's office and immediately start convincing him or her how great it is. You are just about ready, but *you must complete some other steps before you are ready for this.* Your next step is to do some political selling before you enter his office. You now enter a new phase of getting your ideas accepted called "The Politics of Selling Your Ideas" [13].

The first step is trying the play out in Philadelphia before you go to Broadway with it. In other words, you want to test the waters out before you go into your supervisor's office. The best way to do this is by showing the plan to key people in the organization who are respected by your supervisor or to your mentors. In doing so you can also help the saleability of your plan.

Ask their opinion and get a forewarning of how your supervisor may react. They will probably identify the same objections your supervisor will come up with. With this knowledge you have time to develop plans to work around these objections and easily handle them when your supervisor raises them.

If you are a politically savvy engineer, you will attempt to work out any differences between your viewpoint and that of the key people in the organization. You need to realize that the outcome of your effort to sell your plan will be based less on its technical value and more on the fact that enough people want it to succeed. A new idea can be less than perfect technically but still survive if key people involved with it want it to work. Where there is a group will, there is a way, to paraphrase an old saying.

These key people are also in a position to show you how to improve your plan. Ask for their support and any improvements they might have. By asking for their help and support you are effectively getting them to back your plan. They are less likely to shoot down the plan if their ideas are also included. It also provides a check to see if there are any fatal flaws or any major things you have overlooked. It's better to find out that your plan has a major flaw before you go into your supervisor's office than after.

You also quickly find out who supports your idea and who does not. Having the support of key people in your organization adds a large amount of credibility to your plan. Being able to say that the senior engineer backs your plan goes a long way to selling it. Also, being able to respond to the objections of the senior engineer and show how you plan on overcoming them effectively neutralizes the opposition. When it comes time to defend your plan you must know who to call upon for support and what objections other people are using to shoot down your plan.

When you run into people who object to your plan, handle them with caution. First ask them why they are objecting to the idea or plan. Only when you allow them to talk freely are you able to air the differences. Empathize with the people objecting and explore their apprehensions. Get specific information about what exactly they are objecting to. Next, ask them how they would overcome the problems. Sometimes people are objecting because they really want to tell you the solution. Try to get them to help you define a solution. Nothing makes allies quicker than identifying that the senior engineer has identified a serious problem with your idea but he or she is working a solution so your group can realize all the benefits. If you win over resistors, periodically recheck their position to make sure they haven't changed their minds again. This is important.

If you fail to get the support you need don't try to force the situation. Do not argue and counterattack (this is a lose–lose situation). Back off and recognize that you have an honest difference of opinion on the plan. In other words, you agree to disagree about your plan. This allows the objector to state his objections and you to continue; neither of you have lost credibility. By trying to force agreement you may win a short term battle, but the first problem you run into implementing your plan will result in the person running around telling everyone how he or she told you so!

When you present your plan to your supervisor, he or she will probably want to know the reaction of other people. Having already shown it to key people and gotten their support before you go into the supervisor's office allows you to identify who backs it. It also allows you to identify who is against it and the objections that they have, objections that you have shown ways around in your plan. In either case it only adds more credibility to your plan.

Show the plan to your mentors before you go into your supervisor's office. They can give you a good unbiased opinion of the plan. Their years of experience in the company can provide you background on what has gone on in the past, what has previously worked, and how you might best sell your plan.

A word of caution—if your co-workers are open to new ideas, you will find a warm reception for your ideas. But often many workers feel that new ideas create more work, something they are not interested in. To combat this they will claim that things worked in the past and there is no need to change them—"we always did it this way; no reason to change."

Another barrier you may encounter is the "sacred political laws" of your organization, which you may be unknowingly violating. For example, if you are in a DOS-based PC company and recommend changing to a window-based personal computer you may be up against a sacred political law. Chances are some upper level manager has already made the decision and you will be bucking the system.

Remember, new ideas or plans frequently upset the power structures of the organization and call for change. Your plans could be secretly saying, "Hey, supervisor, you are doing everything wrong and here is the right way to do it." Your supervisor will not be very interested in implementing a new plan that only highlights how wrong things were being done. New plans or ideas presented in this light often quickly end up in the trash before they can do any harm to people's careers.

After you have tried your plan out on other people, you should have a good idea of its strengths and weakness. Polish the plan with what you have learned. Incorporate new ideas when you have picked them up and be sure to give credit to those people. Revise your approach if it needs it. Although it may be hard to change things, remember that your end objective is to benefit the company and its customers.

Here is a quick checklist of some of the helpful things you might have found out by sharing the plan with other people.

1. What are the weakness and strengths of the plan?
2. Is the timing or political climate right for my idea?
3. Are there any other issues that I can piggyback on to help sell my idea?
4. Who will make the final decision, and by what means? Committee vote? Supervisor approval? Manager-level approach?
5. Can I call upon other outside sources to help sell my plan? Co-workers? Consultants? Company publications? Technical journals? Competitors?

After you have shared your plan with other people, received their support and backing, and, finally, modified and improved the plan, it is then time to share your plan with your supervisor.

MAKING THE BEST PRESENTATION YOU CAN

Making the big presentation to your supervisor or management may be one of the hardest things you will ever do. You will be nervous the first few times and you must practice several times before you actually do it. Practice giving your plan several times to other people first, if possible. Run through it from beginning to end just the way you intend to give it. Your mentor would be a good person to practice your pitch on. After such practice you should be ready to present your ideas.

Some people will tell you that you do not have to practice, just go in there and present it. However, I feel this is the worst thing you could do. All athletes practice many hours before they actually play a game. Musicians practice hours before they put on a concert, then why not you? Let your supervisor see you at your best, not at your worst.

Make sure the setting and time of your presentation is good for your supervisor. The best way to do this is to talk to him or her in advance and block out a portion of time where he or she can give you some undivided attention with no interruptions. Announce that you have a new idea you would like to discuss with him or her and you need a few minutes of his or her time. Generally keep the meeting to approximately 45 minutes or less.

When the time comes make sure you will not be interrupted. Ask the secretary to hold all calls during your meeting if possible. If the supervisor answers his own calls, then ask if he would mind going to a conference room where you can shut the door. Your objective is to get undivided attention so that the value of your plan can clearly be perceived.

The style and type of presentation is best left up to your judgment [14,15]. It may be as simple as a few typed pages or a complete report with a viewgraph presentation to highlight the important parts. A really effective engineer can personally produce computer spreadsheets, graphs, and viewgraphs for all types of presentations. If you need more training in this area, get it immediately. Poor viewgraphs can ruin any presentation. There is one very important concept that you must remember any time you are making a presentation:

Your technical ability will be judged by your presentation skills and the quality of your viewgraphs.

This is unfortunate since your presentation skills do not necessarily reflect your technical ability. Nevertheless, people do react this way at nearly every presentation. Therefore you must put on the best show you can.

Regardless of the size of the report or presentation, make sure that your supervisor has a copy of it when the meeting is over. Invite her to make notes on it and keep it for future references.

As you present the plan make sure you stop and get feedback. If you're not getting any, ask people what they think. If they like it, you've succeeded. If they do not like it, find out why. Discsuss objections but do not try to override or counterattack. You can always ask what it would take to make it a better plan. If they start to suggest things, you've got them hooked. Try to get everyone to buy into the plan and start supporting it.

If management likes your plan then its on to the next phase: "How do we go about implementing the plan?" Remember, coming up with the good idea is only half of the solution. Implementing the plan and realizing the benefits is the other half. Only after you have started to realize the benefits will it positively affect your career. You must get your supervisor to share with you the details of how the company will approve and implement the plan. Once management has decided to move, only then can you start the implementation of your plan in earnest.

A FINAL WORD OF CAUTION

Here's one final word of caution about selling your ideas. Remember that some people will think only of reasons why it will not work. You must have a strong will to overcome these objections and keep pressing on. Often in a meeting it can only take one no vote to kill a great idea even if ten people have voted yes already. Don't let one "no" vote ruin your plan, make sure it does not have a fatal flaw, then press on.

Finally, most people will say "If I have to go through all this trouble just to suggest something, then forget it." What they do not realize is the tremendous learning experience you go through doing all the tasks mentioned in this chapter. Maybe your first idea will not hit the big jackpot, but perhaps the next idea you will be better prepared and more capable of selling it. If your plan does not work out you still can benefit from the experience. By putting the plan together you are showing your supervisor how you are trying to improve. If it is well organized and neat you are clearly demonstrating your ability to think through a problem in a well organized and methodical manner. These are very valuable skills in an employee.

As a result of your efforts, your supervisor will probably be more willing to have you help on the next big cost saving or cost cutting activity. These are the types of activities that help career advancement. As a consequence, if you follow the suggestions in this chapter you stand a better chance at career development regardless of whether your idea or plan is accepted or rejected. With this type of outcome, what are you waiting for? You have nothing to lose by suggesting something and stand a very good chance of career development.

SUMMARY

Just having a great idea is not good enough. You must have the skills to sell it to your co-workers and supervisor. The first step in selling your ideas is to get organized and write it down. Document the idea or plan. Part of documenting the idea includes doing some research into the good and bad points of the plan, what has been done before, and what benefits you can expect from implementing the plan. Next you must write up the plan in an organized fashion. Discuss it with your co-workers and get their feedback and support. After you have obtained some backing and incorporated any changes or improvements, it's time to make the big presentation. However, practice first. Then give it your best shot under the best circumstances.

ASSIGNMENT

1. Brainstorm for a few minutes and try to think of some ideas that you could implement to save your company some money.
2. Pick the best one and do some research on it.
3. Generate a plan or write-up that describes your idea and its potential benefits.
4. Share the plan with one or more of your co-workers and get their input.
5. Modify your plan and incorporate any good suggestions.
6. Present the plan to your supervisor.
7. Implement the plan.

CHAPTER 11

GETTING A PROMOTION WHETHER YOUR PROJECT IS A SUCCESS OR A FAILURE

During your engineering career you will encounter many so-called "truths" (actually fallacies) about engineering. If you blindly accept these so-called truths you will be severely limiting your career. One of these so-called truths is the belief that you can't get promoted if your project is a failure. The fact of the matter is that most promotions are the result of outstanding performance during an unsuccessful project. The reason for this is that during an unsuccessful project there is usually quite a bit of upper-management visibility. This extra visibility provides a unique opportunity for the engineer to demonstrate his or her capabilities to solve problems and show readiness for promotion. With outstanding performance and this extra visibility all the ingredients are there for a quicker promotion. Unfortunately, the converse is also true—all the ingredients are right for a quicker demotion if your performance is poor.

Another common fallacy is that most engineering projects are successful. The fact of the matter is that 4 out of 5 engineering projects are *not* successful and often end up going nowhere. Therefore you stand a very good chance of spending most of your career working on unsuccessful projects. Another fallacy is that a successful project automatically results in a promotion.

If you blindly accept these fallacies, you have doomed yourself to a very unsuccessful career. To overcome these fallacies and maintain career development you must develop talents or skills that lead to career development regardless of whether or not your project is a success. If the project you are working on turns out to be a failure, the question then becomes "How do I best survive and, hopefully, end up with a promotion?" If the project is a success the question becomes "How do I best capitalize on the results to increase my chances for getting a promotion?" The specific actions you will take depend upon whether or not the project will be judged as a success or failure. In this chapter we will identify specific actions you should take when the project is unsuccessful and another set when the project is successful.

Unfortunately, it is not always easy to determine whether or not a project is a success or a failure. What one supervisor considers success another might consider failure. The first step is to determine if the project you are working on is heading for success or failure. To best do this let us identify some common indicators that will help you judge.

HOW TO DETERMINE IF YOUR PROJECT IS A SUCCESS OR A FAILURE

One indicator that is always good to check on is your supervisor's opinion of how the project is going. When you have a few minutes alone simply ask how the project you are working on is going. Listen carefully to the answer; it will tell you a lot. If your supervisor responds positively then this is a good indication that things are moving along quite well. Try and get him to identify the specific things he sees as significant. This information is very valuable and you may be able to capitalize on it later.

If your supervisor responds negatively to the progress or results of the project, this is a good indication that the project is not turning out as planned. In this case ask what is going wrong and what should be done. Try and get him to identify the specific actions he thinks you should take to turn things around. This is very valuable information.

A word of caution about overreacting to your supervisor's opinion. Some supervisors will claim a project is headed for disaster right up to the day it is successfully completed. This makes them appear as though they snatched victory from the jaws of defeat. Other supervisors may claim success right up to the end when the project ends in total disaster. Supervisors do this to keep all the problems under the rug in hopes that something may come along and save them.

Another indicator is the project technical results. Is the project or design meeting the requirements or is it failing tests? Compare the test results to the original requirements and modeling done. Is it coming close to the results expected or is it orders of magnitude off? Check the results of other areas of the project. Your area may be successful but if another area is failing miserably it could make the entire project fail. For example, a new car body design might look great, but there may be severe problems with the motor and braking system that can cause it to fail. If you are only working on the body design, you may not be aware of the failures in other departments that are hurting the overall success of the project. Overall test results are good indicators of a project's success.

Check the schedule for the project. Is the project progressing as planned? Is it ahead of schedule or behind? Being behind schedule indicates problems. Another indicator is cost. Are the costs running as planned or is the project severely overrun? Overrun projects are bad news and often get the president of the company's attention since they eat up profits.

Check the follow-up plans for the project. Is the project going to transition into the next phase? Is funding available to continue the effort next year? If so, the chances are the project is getting the results expected and it will be considered a success. If not, you may be working on a dead end project. Management may intend to end the project since the results everyone hoped for are not happening. By knowing the follow-up plans you can get a good indication of whether the project is considered a success or failure.

How are management reviews of the project going? If they are smooth and you don't hear much, then it is a good indication that the project is successfully progressing. If the management reviews result in more meetings and sudden changes in the direction of the project, then it is a good indication the project is headed for trouble.

Another good indicator of the success of a project is the customer's reaction. Does the customer seem pleased with the results to date or quite upset?

The health or success of a project will change from day to day and week to week. As the project proceeds, unexpected problems will arise from time to time. Hopefully they will all be solved. Therefore it is impossible to determine whether a project will be successful on any given day. In order to get a good indication of how successful the project will be, you must be constantly monitoring all aspects of it continually throughout the project. Only by doing this will you get a good sense of whether the project is going to be judged a success or a failure.

WHAT TO DO IF THE PROJECT IS A FAILURE

If the project you are working on is headed for disaster you must take actions to minimize the damage to your career. These actions will be in response to the two most common questions asked by management when a project is headed for failure. These questions are: Is the project failing because the wrong people are assigned to it or are the people working on it incompetent? If the answer is no, then the next logical question becomes: Is the project failing because the technical problems are insurmountable? As management searches for the answer to these questions you must be prepared to take action to minimize the damage to your career and hopefully use the failure to advance your career.

Your first response to a failing project should be to go into a high energy and high output state. This means simply making your efforts visible to management to show how hard you are working on the problem. These visible efforts include working extra hours, which means nights and weekends, getting organized, making excellent technical presentations on the problems, identifying solutions, and projecting an attitude of always willing to try. Let's explore some specific actions you can do for each of these efforts.

The more time you put in the better you look and the more you benefit. However, if you put in extra time and your supervisor does not know that you are, you are setting yourself up for disappointment. It is a good thing to first approach your supervisor and ask about working overtime to help out. At the end of the week it is a good idea to stop by the office and let her know what you accomplished and how much overtime you worked. If you plan on working over the weekend, it helps to point this out.

When you are working late at the office and you have voice mail or electronic mail, it's a good idea to leave your supervisor a message just before you go home. Both voice mail and electronic mail have a time stamp indicating the time you sent the message. It will impress the boss when she reads your message or progress report the next day and it indicates that you sent it at 10:30 PM. Also, a message sent on the weekend so she can be up to speed on Monday morning is another way to highlight how much effort you are putting in.

The next thing you must absolutely do is get organized. This may be hard to do when things are falling apart all around you, but you must! Remember, management is constantly asking "is the project failing because of the people assigned to it?" If you present an image of being organized, the answer to this questions will be "no!" In fact, management will probably walk away saying "thank goodness we have her on the project; if anyone can turn this around it will be her!" These types of statements ultimately result in career growth and promotions.

Now the question becomes one of how can you show that you are organized? It is simple. The first thing to do is generate a plan. What tasks are you going to be doing to solve the problem? What is the schedule for accomplishing the tasks and what results do you expect? The best way to communicate this to management is through a written plan that documents your intentions. By writing up a plan and giving it to management you are accomplishing two things. First you are showing them that you are organized and second you are providing them with a "get well plan" that they can share with the supervisors. If the plan is good, upper-level management will get the impression that your supervisor has assigned the right person to solve the problem, making him and you look good.

Next you must complete a thorough technical analysis. First identify the specific technical problems that must be overcome. Next identify potential solutions to the problems. Determine the good and bad points of each solution. Next rank the solutions and, finally, present a recommended approach. The worst thing you can do is present all the problems to upper-level management with no solutions. An engineer is paid to understand the problems and identify solutions. Do your job if you expect to be promoted! Even a project that is considered to be a failure will be looked upon as a success if you can explain exactly what went wrong and how to fix it.

There are two conditions for success as far as your career is concerned.

1. The project accomplished its goals and was successful.
2. The project encountered severe problems and failed, but we know exactly what went wrong and how to fix it next time.

In either case the project results should be good for your career. There is only one condition for failure: The project failed and no one knows why.

When a project is heading for failure there often are a series of meetings with management to ensure that everything possible is being done to make the project successful. If you are invited to one of these meetings never go unprepared. Make sure that you have a well organized and thought-out report. We have already discussed the three major things you must bring to all the meetings; a plan, potential solutions, and good technical analysis. A well organized and neatly prepared handout summarizing all three areas will be appreciated.

Management meetings will be tough and very stressful. It is important that you be well prepared and have thought through your ideas. Your plan should identify key tasks and dates. It should show a logical sequence of events that you plan on following. Take the time to explain the importance of each task and how the task will contribute to the solution. Make sure the plan is realistic and can be accomplished within the time you have budgeted and with the resources available. If you need more time or resources then identify the need.

The technical part of your presentation should describe the problems. It must be exact and any analysis presented must be correct. Use graphs and math modeling supported by any test results you have obtained. Compare modeling results to test results as much as you can. A thorough review of test results is always looked upon very favorably.

Remember your objective is to demonstrate that you have the technical knowledge necessary to successfully solve the problems and you are the right person for the job. Bring photos of anything that will help you illustrate the problem. The optimum career move is to bring the managers down to the lab and let them see, handle, or run whatever it is you are working on. People are more sympathetic when they see first-hand how difficult the problem is and will naturally become involved.

The attitude that you project at review meetings and during discussions with your co-workers also affects career development. The project may end in failure, but with a good attitude you can minimize the damage that failure can do to your career. What you would like to hear your supervisor say at completion of the project is "The project was a failure but you had such a good attitude that next time I want you on my team again!"

I strongly recommend not to take things personally and start fighting during meetings. This may be hard to do as people start to criticize your plans. You need a calm and level-headed approach when everyone else around you may be losing their composure. Rather than arguing with them, spend the time and energy in drawing out their ideas on how to make improvements. This will get their ownership in the plan and they will be less likely to blame you when their ideas don't work either.

One important philosophy to adopt is:

> Be part of the solution and not part of the problem.

This attitude is one of always looking for solutions and volunteering help. Supervisors usually welcome someone willing to take on more work or try out something new after hours. Be willing to try out new ideas even though they are not yours. Often someone else will have a potential idea and need you to try it. Be willing to give it a try even though you may not agree with it. Sometime just following orders can benefit your career more than you realize.

Stay away from any negative statements. Being Ms. or Mr. doom and gloom does not help the team and may even contribute more to the failure of the project. Some of the doom and gloom statements that indicate to management that they have the wrong people on the team are:

That will never work because. . . .
There is nothing we can do. . . .
It's impossible. . . .
No sense trying, the project is a failure because. . . .
Why won't it work? (As compared to: What will work?)
It's not my idea, no way am I. . . .
It's not my fault, it's their fault for. . . .
It's not my department that. . . .

Some of the good-attitude statements that reinforce managements conviction that they have the best person on the job are:

I'd be willing to try that because. . . .
The good points of the solution are . . . ?
I think I might be able to make it work if we. . . .
I'd be happy to put in extra hours to see if it works.
How can I help the team out?
What are the good points of the. . . .
It doesn't matter who cause the problem, we must. . . .
Just tell me what I can do, I want to help with. . . .

The reason that attitude is so important is that most managers have worked on projects that have failed. They realize that tough times require a positive attitude. It is the person who keeps on going who will eventually succeed, and attitude has a lot to do with it. Besides, all projects eventually come to an end and there will be the next project to work on. Supervisors will make assignment for the next project based on the performance demonstrated on the last project. Who do you think they will choose, a person who projects doom and gloom or one who has a positive attitude and is willing to work through problems. Who do you think they are going to promote regardless of how the project turns out?

At the end of an unsuccessful project one good action is to document the lessons learned. This may take the form of writing a simple memo that documents all the good and bad lessons learned during the project. Volunteering to write this memo is good for several reasons. First, management looks on this activity very favorably since it helps share with other groups those things that worked so they can capitalize on them and avoid those things that failed. Second, after you have written several of these memos you will have acquired an excellent library of things to do and not to do on a project. This knowledge is power for future projects.

Finally, do not fix blame on any one individual. It was a team effort and everyone failed together. The best thing you can do is learn from your mistakes and move on. Most people can handle success, but the really successful people in life are those who can learn to handle failure and recover.

WHAT TO DO IF THE PROJECT IS A SUCCESS

The discussion thus far has focused on what to do if the project you are working on is headed for failure. Now let's look at what to do if the project you are working on is headed for success.

Most people believe that project success guarantees a promotion. This could not be farther from the truth. Project success does not guarantee a promotion, but it does help. The reason that success does not guarantee a promotion is due to the fact that management often will ignore successful projects and instead devote their time and energy to projects that are in trouble. Another common response by management is "Why should I promote you for a successful project? That's what you get paid to do. You are just doing the job I hired you to do in the first place. If it weren't a success you wouldn't be doing your job. We don't give promotions for just doing your job. You must therefore take actions that maximize the benefits from working on a successful project.

As with failure, your response to a successful project should be to go into a high energy and high output state. This means simply making your efforts visible to management to show the excellent results you are obtaining. Again, these visible efforts include: working extra hours, which means nights and weekends, getting organized, making excellent technical presentations on the great results, identifying benefits, and projecting an attitude of success.

Do all these actions sound familiar? They should because they are the same actions you would be taking for a project that is failing, but with a different twist on them. Let's explore some specific actions you can take for a successful project.

The more time you put in the better you look, and the more benefit gained. If you put in extra time and your supervisor does not know about it or he does not know how well things are going, you are setting yourself up for disappointment. At the end of the week it is a good idea to stop by his office and let him know how much overtime you worked and the exciting results you obtained. Again, the same efforts the failing project required.

The next thing you must absolutely do is get organized. The question is how can you show that you are organized? Again, it's simple. The first thing to do is generate a report showing all the good results. Highlight the things that went well. Compare modeling results to test results. Show how the results met or exceeded plans. By writing up the results and giving them to your boss you are accomplishing two things. First, you are showing that you are organized and, second, you are providing a "Good News Report" that can be shared with other supervisors. Upper-level management will get the impression that your supervisor has assigned the right person to the job, making your supervisor and you look good.

Often times when a project is successful management feels no need to review progress. In this case, you call the meeting! Make sure that you have a well organized and thought-out report. The report should summarize the plan followed, any technical analysis, test results, and a benefits summary. A well organized and neatly prepared handout summarizing all three areas will be appreciated.

The technical part of your presentation should describe the problems you solved. It must be exact and any analysis presented must be correct. Use graphs and math modeling supported by any test results you have obtained. Compare modeling results to test results as much as you can. A thorough review of test results is always looked upon very favorably. Remember, your objective is to demonstrate that you have the technical knowledge that contributed to the success of the project and you are the right person for the job.

Bring photos of anything that will help you show the success. The optimum career move is to bring the managers down to the lab and let them see, handle, or run whatever it is you are working on. People are more likely to appreciate the trouble you went through and the magnitude of the success when they see them first-hand.

The attitude that you project at review meetings and during discussions with your co-workers is also important to career development. Make sure that you give credit for success and report things in terms of "we" and not "I." The reason this attitude is so important is that most managers have worked on projects that were successful. They realize it required a team effort and no one individual did it all. Supervisors will make arrangements for the next project based on the performance demonstrated on the last project. Who do you think they will choose, a person who takes all the credit or one who is a good team player and willing to share the credit. Who do you think they are going to promote?

If you are a team leader for the project there are several additional things you can do to help the team capitalize on the results. First you can nominate the team for a company award. Or you might try to get an article published in the company newspaper. Make sure the team members get their names in the article. Another good thing to do is take a team photo and pass out copies to the team members.

Photographs of the hardware or the test results are always a good thing to give your supervisor to show others or put them on an office wall. If your supervisor hesitates about doing anything, you might point out the benefits received when it becomes known that he was responsible for assembling the team.

If it is possible, you may want to publicize the good results outside of the company. Writing a paper and submitting it for publication or presentation at a symposium is an excellent idea. This gets your name known throughout the industry rather than just throughout the company.

At the end of a successful project, it is again a good idea to document the lessons learned. This may take the form of writing a simple memo that documents all the good and bad lessons learned during the project. Volunteering to write this memo is good for several reasons. First, management looks on this activity very favorably since it helps them share your work with other groups so they can capitalize on it and avoid those things that failed. Second, after you have written several of these memos you will now have acquired an excellent library of things to do and not to do on a project. This knowledge is power for future projects.

SUMMARY

The so-called "truths" that you encounter in your engineering career should not be accepted at face value or they limit your career growth. This is especially true of the fallacies that successful projects always result in promotions and projects that fail will limit or damage your career. There are actions you can take to ensure that your career continues to grow regardless of the outcome of the project. Hopefully, you have realized that the actions are similar for the successful project or the unsuccessful project.

For an unsuccessful project, you need to put in extra effort, get organized, develop a recovery plan, brief management, and identify the reasons for failure. For a successful plan, you also need to put in extra effort, organize the good results, brief management, and identify the reasons for the success. In either case you will be operating in a high energy state and a positive attitude, giving credit where credit is due and not fixing blame. Excellent technical presentations showing theory, modeling, and test results are musts following both failed and successful projects.

ASSIGNMENT

1. If you are working on a major project, determine if management considers the results to date to be successful or unsuccessful.
2. If it is successful, what should you be doing?
3. If it is unsuccessful, what should you be doing?
4. Can you name any other fallacies that might be limiting your career that you are not even aware of? For example:

No raises or promotions are given in bad economic times.
They won't promote me because I'm. . . . They never have before!
(Hint: Fallacies are usually great sounding reasons that put blame on some abstract or uncontrollable circumstance.)

CHAPTER 12

THE VALUE OF VISIBILITY—HOW TO GET IT AND USE IT

We all desire to have our good work recognized and appreciated. Nobody likes to have their hard work ignored or downplayed. Every person fights for visibility or recognition—it is a natural human instinct. However, most engineers consider this subject taboo and fail to realize the great importance that visibility plays in their career development. In this chapter we will address this taboo subject and hopefully make it something you will feel more comfortable discussing and using.

Good visibility can have a very positive effect on your career advancement. Correspondingly, poor visibility can have a very negative effect on your career. Knowing the benefits of good visibility as well as the dangers of poor visibility is essential for career advancement. Therefore we will first discuss the value of good visibility and then the dangers of poor visibility.

Next we define what exactly is visibility for engineers. The engineer must be aware that there are two types—technical and social visibility. Definitions and examples of these visibilities are presented. Following this, we discuss how to develop and use your visibility. And, finally, guidelines are recommended to ensure that you use your visibility correctly.

UNDERSTANDING THE VALUE OF GOOD VISIBILITY

The value of good visibility is accelerated career advancement. If you have good visibility with the upper management of your company and co-workers, you significantly enhance your chances for career advancement. Without visibility, you significantly reduce the chances for career growth.

Good visibility with the upper levels of management is absolutely essential. The upper levels of management are where most of the decisions will be made that will directly affect your career—the projects you work on, who gets raises and promotions, and who will be laid off during economic downturns or cutbacks. Upper-level managers will promote people they know and people who have demonstrated good performance. Please note the key words in the last sentence, "know" and "demonstrated performance." Visibility is the means to obtain these. Upper-level managers can only get to know you and your performance if you have visibility with them.

It is easier for managers to lay off nonperforming people they do not know than people they personally know who have demonstrated good performance. Therefore, good visibility with upper management contributes to your advancement and keeps you off the layoff lists.

Good visibility with upper management provides other benefits for your career. By developing and using your visibility you will have the opportunity to get the recognition that you and your team deserve. This visibility may result in team awards and bonuses. If you and the team have done a good job, why not get the credit? Good visibility with management provides you with the opportunity to show off your good work.

Visibility with upper management can also provide you with unique opportunities that are not normally available to most engineers in the company, such as getting the better assignments, or being considered for those special projects. It also allows you to have the attention of the decision makers. This attention is something that is extremely important if you need help or additional resources committed to your project to make it successful.

Developing and using your visibility with management can also help you with your co-workers. When it becomes evident to your co-workers that you have good visibility, they may start coming to you for help and support. If co-workers know they can get management visibility simply by working with you or associating with you, they will be more likely to help you out or join your team. Co-workers might also want you on their projects, since they realize that having you on the team will result in upper-management visibility for their project.

Good visibility with your co-workers is also very beneficial to your career. It means that you are well known (highly visible throughout the company) and your opinions are usually sought out. This is beneficial to your career in several ways. First, when a group or co-worker has a problem they cannot solve more than likely they will seek your advice. This makes you a valuable employee. Second, with your opinions being sought out, it is more likely you will work on the more challenging and interesting engineering problems. Third, the more people in the company who know you the less likely you will run out of work and be placed on the layoff lists.

Having good visibility with upper management and your co-workers offers many benefits to your career. To put it simply:

> If you have good visibility and use it well, you have power!

Visibility is a double-edged sword. Good visibility can help you; correspondingly, poor visibility can significantly hurt you. Let's explore how poor visibility can hurt you and what to avoid.

UNDERSTANDING THE DANGERS OF POOR VISIBILITY

Poor or bad visibility is getting management's or your co-workers' attention when you are not able to show them your best. Visibility when you are performing poorly is poor visibility, and it will have a negative effect for your career. It's just what you do not want for career advancement. Poor visibility is usually the result of inadequate planning or incomplete work on your part. Inadequate planning and poor work do not get anyone considered for advancement. In fact, they do just the opposite for your career. Managers think in terms of what projects they can assign the badly performing engineer to so he or she won't get things screwed up.

Poor visibility results in the engineer being labeled as a nonperformer in the eyes of management. These engineers usually get the mundane assignments—simple assignments that are not challenging and do not lead to career advancement. They seldom get to choose their assignments and are usually passed over for raises and promotions.

Visibility with upper management does not occur often, so when the opportunity occurs, you must work to ensure you are at your best when you get it. Please put in the extra effort to give the best image of your work you possibly can. It will be well worth the effort. As one senior engineer once told me,

> Management judges you on performance you have demonstrated. You judge yourself on what you are capable of doing. Usually there is a big difference!

If you have the opportunity to get upper management's visibility for your project, remember that the managers you will be briefing are continually getting briefed by other people in the organization. They have seen thousands of presentations and work well done. They will quickly recognize inferior or incomplete work. So if you plan on using your visibility with upper management to show off your accomplishments make sure they see you at your best.

Bad visibility also occurs when you are seen in a way that makes other team members jealous or feel left out. These feelings usually result when "only you" present the results to upper management, or "only you" do the talking, or you hog all the credit by using the term "I" instead of "we." This type of bad visibility can result in your alienation by peers. Alienation by your peers will ultimately result in your failure because you cannot do it all—you need them.

Poor visibility also occurs when you use your visibility to spread bad news up the chain. If you are continually using your visibility to spread bad news, you develop the image of a troublemaker—just what you don't want for career development. People will immediately think that trouble is coming whenever they see you. As one manager once said to me when we observed another engineer approaching, "Here comes trouble. Every time I see him it's bad news." With this type of image you usually are not the first picked for the high profile and technically challenge assignments. Managers want engineers who are always talking about solutions to problems, not doomsday prophets.

Another reason not to use your visibility to spread bad news is the "shoot the messenger syndrome" that many mangers have. In this syndrome the manager simply blames the bad news on the person who brings it, regardless of whether they are responsible for it or not. Once you have been shot for bad news, it will quickly cure you of using your visibility this way. I have personally learned this lesson the hard way. Junior engineers need to be especially careful of this since many senior engineers will gladly let the junior engineers get management visibility when it comes to bad news. The junior engineer is all excited to finally get to brief the boss and rushes in with the bad news only to get scolded.

Poor visibility also occurs with your co-workers when you produce bad work that ends up making the team look bad. For instance, you may come up with a new solution to a difficult problem. In your excitement, you quickly spread the good news to the team before you really have had time to check it out thoroughly and make sure it is a good solution. Then, after you have had time to work on it more, you discover that it will not work. Now you must go around to the team and admit your mistake. This is not the kind of visibility that you want. If you let this happen more than once, it is highly probable that the team will quickly learn to ignore you, something you do not want for career advancement.

So far we have identified the benefit good visibility has on your career, namely, accelerated career advancement. And we have also identified how bad visibility can significantly limit your career. Now let us shift our attention to exactly what is visibility for engineers, how you develop it, and how to utilize it for career advancement.

DEFINING TECHNICAL AND SOCIAL VISIBILITY FOR THE ENGINEER

Each person perceives and interprets the meaning of visibility differently. Their definitions of visibility are derived from their personal experiences. Therefore, we need to come up with a common definition of visibility. The dictionary defines visibility as "capable of being seen, evident, at hand, or available for observing." This definition does not provide much insight into what visibility means for the engineer.

For our purposes, let's expand this definition. Let us define visibility for engineers to mean "making management and co-workers aware of the engineer's skills and accomplishments for the purposes of benefiting the company, one's own career, and those of others." With this definition in mind, let's further quantify visibility into "technical visibility" and "social visibility."

Technical visibility is making management and co-workers aware of the engineer's scientific skills and technical accomplishments. Examples of technical visibility are writing technical reports, demonstrating how hardware works, reporting and explaining new scientific methods, filing for patents, and publishing papers, to name just a few. If your career plans are to become a staff engineer or a great technical expert, then technical visibility is your primary concern.

Social visibility is making management and co-workers aware of the engineer's social interaction skills, communication skills, and team leadership capabilities. Examples of social visibility are behavior in crowds, self-image or self-appearance, team leadership, progress reporting, resolution of team/customer conflicts, and presentation or speaking skills. If your career plans are to become a manager and, hopefully, the vice-president of the company someday, then social visibility is your primary concern.

Before we start discussing technical and social visibility, I'd make to make an observation about developing and using these two visibilities. Most great technical experts spend little time on social visibility. Their claim is that great technical work speaks for itself and therefore they develop little social skill and visibility. As a result, many people find technical experts hard to be around and avoid them as much as possible. The technical experts feel their work is unappreciated and often end up in shouting matches with technically incompetent managers who do not understand. By not developing and using social visibility, their technical achievements go unnoticed and their career development is often limited.

Correspondingly, I have seen managers who have developed nothing but social visibility and no technical visibility. These managers quickly rise to the top but fail to remain there. Subordinates quickly realize there is no technical depth to these managers and find it hard to respect them and their decisions. Not developing and using technical visibility ultimately results in failure to successfully lead the group and ultimately ends in replacement.

The point that I'm making is that to be truly successful you must have both technical and social (T&S) visibility in your career. I have observed successful technical experts and managers who have both. They must have both to advance and survive. The key is to have the correct balance for the career path that you choose.

You must start to develop your T&S visibility from the day you leave school until the day you retire. Developing and using both technical and social visibility is a never-ending career activity. I have identified three areas or domains where the engineer must develop T&S visibility. These areas are identified in Figure 12-1 as : 1) local visibility within the engineering group or department; 2) company-wide visibility; and, finally, 3) outside-the-company visibility.

The engineer needs to develop T&S visibility within the group or department first. This domain is limited to the engineer's immediate co-

DOMAIN	TECHNICAL VISIBILITY		SOCIAL VISIBILITY	
WITHIN GROUP OR DEPARTMENT	ATTEND COURSES WRITTEN REPORTS DESIGN TRADEOFF REPORTS HARDWARE DEMONSTRATIONS ANALYSIS REPORTS ATTEND CDR, PDR	SIMULATIONS VIDEOTAPES OFFICE ARRANGEMENT PHOTOGRAPHS MODELS RESOURCE LIBRARY	GROUP LUNCHES SOCIAL MEETINGS TEAM SPORTS- GOLF, SOFTBALL, VOLLEYBALL LAB LOCATION	DESK LOCATION SEAT LOCATION IN MEETING
COMPANY-WIDE	FILE FOR PATENT SUBMIT TEAM FOR AWARD OFFER TO TEACH CLASS WRITE ARTICLES FOR COMPANY PAPER PRESENT AT CUSTOMER REVIEWS		VOLUNTEER FOR COMMITTEE BLOOD DRIVE UNITED WAY/ SAVINGS BONDS DRIVES COMPANY PICNIC	
INDUSTRY-WIDE	TEACH A COURSE PRESENT PAPER AT CONFERENCE PUBLISH ARTICLES IN TRADE JOURNALS CONTACTS AT UNIVERSITIES USE OF CONSULTANTS WRITE A BOOK		COMMITTEE MEMBER OF SOCIETY SOCIALIZING AT CONFERENCE HOSTING A SOCIETY MEETING AT YOUR COMPANY	

FIGURE 12-1 Methods of developing technical and social visibility.

workers and managers with whom the engineer is interacting on a daily basis. As the experience base grows and the career advances, the engineer needs to expand visibility beyond the group. The engineer needs to get T&S visibility across the entire company, which is the second domain. The third and final domain is the largest and encompasses the industry the engineer is working in, which means having national and possibly international visibility. Now let's explore some means of developing and using T&S visibility in all these domains to benefit your career.

DEVELOPING AND USING TECHNICAL VISIBILITY TO BENEFIT YOUR CAREER

The primary reason for developing technical visibility within your group or department is to demonstrate to the people immediately around you that you have the technical skills to handle your present job. It is also to show that you possess the additional skills and talents necessary to handle larger and more technically challenging assignments—assignments that will ultimately lead to your advancement when they are successfully completed.

When you obtain or use visibility you must do it in a caring manner that benefits not only you but your co-workers and the company. Only by using visibility in this manner will you truly help your career. Getting visibility only for yourself is not good visibility and not good for your career. The following examples highlight some win–win visibility activities. Let's start with your office or desk area.

How do you have your office arranged? Does it subconsciously say to everyone who visits your office that you are technically competent? Here are some ways to get technical visibility within your office or cubicle.

Do you have a computer on your desk? What is on the screen most of the time? If it is a screen saver that utilizes a cartoon, that says one thing. If it is a complex three dimensional color graph showing the latest results of a new and improved software package that you obtained, that says another. How about your office or cubicle walls? Put up diagrams of the models you are utilizing—the more complex the better. How about plots showing your latest analysis results? How about photographs of the hardware you have worked on? Put up any technical awards or patents or plaques that you have received. Stuffing those into the drawers does nothing for your technical visibility. When people come to see you they get an instant snapshot of what you are working on. These little snapshots can also help spread knowledge around the company. People often will remember what you were working on and send people to you for help. Or they spread word about software packages. If they are having problems with this application and see the display on your computer, it indicates that you certainly know how to use it. Good technical visibility is good for you and the company at the same time.

How about reference books? Do you have a bookcase with reference books in it? Perhaps your old school books? Expand and fill your bookcase with new reference books. The company library should have many that you can check out and read. Put them in plain view so everyone who comes into your office can see all the good technical resource materials you have.

You can have technical visibility with the boss by simply ordering new books for your resource library. Most supervisors have a budget for this. Why not get the company to expand your resource library and have them pay for it? The visibility you get when you ask your supervisor's permission to order a new book on what you are working on is great. When the book arrives makes sure you show it to your supervisor and explain the things you have learned from it. Show the book to your co-workers and let them know where they can find it. The technical visibility is great. If you do this several times with several different books, you will soon have people seeking you out to use your resource library—just the technical visibility you want. It's even better if you can show them where in the book to find the information they need, thus saving them time. This is good technical visibility that helps your career and others at the same time.

Another good thing to have in your resource library is videotapes. Often vendors will supply you with videotapes showing the benefits of their products. This is excellent resource material and also provides good technical visibility for you. Universities often sell excellent technical tapes; these are also great for your reference library. Remember that the tapes are not useful sitting on the shelf; they only benefit you when you use them. Share them. Universities also loan out tapes; check with your local library and show them to our co-workers. This is also excellent technical visibility.

The next place you may want technical visibility is in the lab. What does your lab area look like? It should be a well organized and neat scientific lab with well thought-out experiments. Put up plots of the data you have collected. Organize the experiment so your boss can observe the results as they are being collected. Drag him or her into the lab if you have to. If your boss claims no have time to see your technical accomplishments, reserve the time with your boss. Stop by and see the secretary to schedule an appointment. Be aggressive about reserving time on his schedule. I had a supervisor emphatically tell me that he had no time to visit the lab and see the team's hard work. I would not take no for an answer and reserved time on his schedule for a quick lab demonstration. When the time came I had to literally drag him by the tie down to the lab. After he witnessed the team's accomplishments in person, he got so excited he proceeded to drag his supervisor to the lab. The two of them played for hours. As a result the project got the additional funding it needed. It was great technical visibility.

If all else fails and you cannot get your supervisor to the lab, then bring it to her. I don't mean move all the lab equipment to her office, but bring all the test results, graphs, photos of the setup, computer analysis, and anything else you can carry into her office. Then go through all of it, piece by piece, until you have shown all the great technical work you have done. In any case, make sure you and the team receive the technical visibility your fine lab work deserves.

When your lab work or assignment comes to an end, how do you get technical visibility? Write a technical report and publish it. There should be previous reports around for you to use as a guide in preparing yours. Make sure the technical content is correct. Fill in with analysis and lab test results. A good outline for your report might be:

1. Executive Summary
2. Theoretical Discussion and Modeling
3. Design and Objective of Experiment
4. Results
5. Conclusion & Recommendations
6. Appendix: Supporting Analysis

Once you have completed the report, take it to your supervisor for review and approval. His or her review and approval of the report is another great opportunity for technical visibility. Perhaps the senior staff people should review the report for approval also—more technical visibility. After you get everyone to approve it, publish it and make copies for everyone in the group. Show your report and distribute it company-wide to other engineering groups—more technical visibility. Make sure your name and those of other members of the team are on the cover. If the report gets distributed throughout the company, other engineers will know where to go to discuss the results, since your name is on the cover.

Some people have indicated to me that putting their name on the cover is hard for them to do. It's like bragging or showing off. To this I reply, it is just getting credit for your hard work. In addition, you give other engineers the opportunity to easily find you. It might be that other engineers are having the same problems you had and by sharing your report and knowledge you will save the company thousands of dollars. Isn't this a good thing? I do not call this bragging, I just call it good engineering practice. It's good technical visibility that benefits you and the company.

Another way to get technical visibility is take a class or attend a seminar and report back to the group on what you have learned. Schedule a meeting, prepare handouts, and briefly summarize the technical highlights that you consider to be significant. Make sure the material you are presenting and summarizing is pertinent to your work. If you feel the seminar does not warrant a special meeting, then issue a written report. In either case, people will be aware of the material and know where to go to get additional technical details. This is especially important if the class is not that worthwhile. You can save the company money by spreading the word.

Technical visibility can also occur through the use of vendors or subcontractors. If you know of several vendors that are developing or have a product that will help your group perform work easier, then invite them to demonstrate their products at your company. Always make sure you have your supervisor's agreement and that he or she can attend the demonstrations. After the demonstrations, write a report comparing the vendors and recommend a way to proceed. Make sure the report does a thorough job comparing the technical benefits of all vendors and their ability to solve the technical problems. After you issue the memo you will become the technical person to contact, resulting in more group technical visibility. This is saving other people in the company time, since you already have seen what the companies have to offer.

Another way to get technical visibility is by videotape reports. Younger engineers have fewer problems with this than older engineers. If a videotape is done with an emphasis on technical content rather than marketing content, this can be a really good means to obtain technical visibility. The videotape must be done in a similar fashion to the written report, with good use of graphics to show theoretical analysis and modeling as well as the results. Only when this is done will the videotape be seen as a good technical report and not just a marketing tool.

Most of the above examples deal with how to develop and use technical visibility in your department or group. Now let's turn our attention on how to get company-wide technical visibility. Since the junior engineer has little opportunity for company-wide visibility, the following examples are more for the senior engineers.

Is the work you are performing patentable? If you don't know, call a meeting with your supervisor to discuss the idea—more technical visibility. If you get the green light to proceed you will have company-wide technical visibility. Most companies have a patent-review committee. The committee is made up of senior staff engineers who review the technical merits of the patent applications. To appear before the patent committee and defend your patent is very good technical visibility.

If you get the go-ahead from the committee to file the patent you will receive further company-wide visibility. Most companies realize that many patents do not get accepted by the U.S. government. They also realize that in order to get committee approval and start the filing process the work must have been very good. Therefore companies often give an award simply for filing the patent. This results in more company-wide visibility since these awards are usually written up in the company newspaper. The conclusion one reaches is if you are doing something unique, try for a patent. You have nothing to lose by doing so and a significant amount of technical visibility to be gained for you and the team, which will benefit all who are involved.

Another way of getting company-wide visibility is to make presentations at any program design reviews involving the customer. The most common and well attended program design reviews are commonly referred to as Preliminary Design Reviews (PDRs) and Critical Design Reviews (CDRs). If there is any way you can talk your supervisor into letting you make a presentation, do it! Usually upper management attend these reviews to show support of the program in front of the customer. If you do a good job presenting your accomplishments you will have excellent visibility. This type of visibility benefits you and the company since your good presentation builds the customer's confidence that the right company has been chosen.

Junior engineers can be frightened of these major program reviews since they will be presenting in front of a large crowd that can be extremely hostile. If you feel that you are not ready to do this, don't panic. You can still get technical visibility and build up your confidence slowly by calling your own meetings. One of the most startling revelations I made as a junior engineer was that I was allowed to call meetings. Simply call a meeting and present what you are working on. Be open to feedback on your work and solicit methods for improving it. This is good visibility for you and at the same time the group gets to review your progress and accomplishments. Progress and accomplishments are something that managers are always looking for.

One means of receiving company-wide technical visibility is to submit the team for an award. This type of visibility is more for the senior engineer or team leader. Since junior engineers are normally not team leaders this applies more to the senior engineer than the junior engineer.

The first step in submitting the team for an award is to write up the technical achievements and meet with your supervisor to discuss the submittal. Most supervisors will immediately ask why these technical results are significant, how well the team worked together, and how will the company benefit. This is excellent technical visibility for you and the team.

Some engineers may feel it is just not worth the effort. They fear their supervisor will never approve the award. However, most supervisors will approve and submit the team for an award. They do this for a couple of reasons. First, the supervisors know they are not going to be very popular with their troops if they turn them down. Second, many supervisors like to draw attention to the work their employees have accomplished—it also helps the supervisor's career. I have highlighted these reasons for a purpose. When a supervisor starts to turn you down for a team award you might mention these reasons why he might reconsider. It could possibly change his mind.

When your supervisor approves, it's on to the next level of approval, which usually involves upper-level managers. At this point you have company-wide technical visibility just for submitting the team for an award. If the upper-level managers do not grant the award you still get some company-wide visibility for the team. If they grant the award, you have gotten the team a reward for their fine technical efforts as well as some company-wide technical visibility.

Another method of getting company-wide technical visibility is to teach a course after hours. Some companies have structured after-hours programs for their employees. If you have taught courses previously this is an excellent means of getting company-wide technical visibility. Often companies will distribute pamphlets listing the courses and the instructors throughout the company. This is excellent visibility. In addition, the students taking the course are from the entire company. Having taught several after-hours courses myself, I soon realized that wherever I went in the company there was always someone there who knew me because they had taken my course. As a result I gained company-wide technical visibility.

The student also received several benefits from me by taking the course. Can you name them? (See the chapter on further education.) The course also provided me with the opportunity to practice and improve my technical presentation skills. For the junior engineer interested in pursuing this, I highly recommend that you team up with a senior engineer and jointly teach a class.

The junior engineer may not want to teach a class, but there are other options available. One of these is inviting local university professors to come and visit. Professors are always interested in seeing the work the company is performing as well as reporting on their work at the university. By arranging for the visit and guest lecture everyone benefits. The junior engineer receives technical visibility and the co-workers get to see the latest research being done at the university, which may be of interest. The professor has the opportunity to present his work and even do some consulting. Everyone benefits from this type of good technical visibility.

Writing a technical article for the company journal is another way of getting company-wide visibility. Many companies publish a technical journal highlighting the technical advances made during the year. A well written and organized paper showing theory and lab results is an excellent way of getting company-wide technical visibility.

As the engineer's career develops and advances, opportunities for technical visibility become available outside the company on a national and even international level. These opportunities become available through the submission of papers to symposiums, engineering society membership, and independent consulting. The value of this visibility internally to the company is that it can claim to have a nationally recognized expert on its staff. Often this helps companies sell contracts. The value to you is that you can point to the fact that you are a nationally known expert at raise and promotion time.

External visibility takes years to develop. It also has what I call a second-order effect on your career. That is to say, internal visibility will benefit your career considerably more than external national visibility. For instance, your supervisor will be more interested in the fact that you finished your work on time and met all the deadlines than in your publishing the results in a paper. Don't get me wrong, external visibility helps your career internally, but not to the extent most people think it will.

Visibility outside your company spreads your name around the industry. This type of external visibility becomes useful if you change employers. Often the people you will be interviewing with read the journals and publications. If your work is good enough to be published, chances are they will have seen it. As part of your interview you can point to your published work to show what you have accomplished.

We have explored various methods of developing technical visibility. However, for advancement the engineer must have also have social visibility. Let's now explore some means of developing and using social visibility for career advancement.

DEVELOPING AND USING SOCIAL VISIBILITY TO BENEFIT YOUR CAREER

As shown in Figure 12-1, social visibility also occurs at the group, the company, and industry-wide levels. The primary reason for developing social visibility within your group or department is to demonstrate to the people immediately around you that you have the social skills and team leadership skills necessary to handle your present job. It is also to show that you possess the additional skills and talents necessary to handle larger and more challenging assignments, perhaps assignments that will ultimately result in your advancement.

Good social visibility is the result of practicing good office manners or etiquette. Nobody likes to be around prima donnas, who are without social skills and offend people every time there is a gathering. Conversely, everyone likes to be around a person who has a pleasant disposition and is polite during a gathering. Good office etiquette is necessary for career advancement.

To gain good social visibility you must do it in a caring manner that not only benefits you but others in the company. If you develop and use your social visibility in this manner it will truly help your career. Getting social visibility only for yourself is not good visibility and not good for your career. The following examples highlight some ways to obtain some social visibility. Let's start with your office or desk area.

Is your desk located in a cubicle along with several others? Can you hear everything they say? Are there many distractions? Are you nearly on top of each other all the time? Being located in this type of situation can be very difficult. In order for you and your office mates to function there are social rules and courtesies you need to extend to one another. For example, when several people come to visit you, leave your bay and go into a conference room. This way you can have a meeting and not disturb your office mates.

Continually talking when your office mates are trying to get work done is also not desirable. Respect their need to concentrate and limit your conversation. Keeping your desk neat and your work organized and not all over your office mates' desks is also good etiquette. Do you ask if you can borrow something before you take it? Do you take the time to get complete phone messages when you answer the phone? These are all factors contributing to your social visibility. If you have bad social visibility, believe me, your co-workers will let your supervisor know. Having bad social visibility with your office mates will hurt your career, not help it—something you do not want to happen.

Check out your social visibility with your office mates—see what they think of you. Sometimes you can be doing things that are offensive to them and may do not even realize it. Be aware of their feelings; it contributes to good social visibility and good work habits for the company.

Social visibility with management can also occur just by the location of your desk in the company. The optimum place you want your desk to be located is in just within hearing distance but not directly in the line of site of your supervisor's office. This location provides you with excellent social visibility. Let's look at some of the benefits.

On any given day your supervisor will come and go from the office many times to attend meetings, check on progress, go to lunch, etc. If your desk is close to the paths used, you will have many opportunities to talk with your supervisor and be seen. This provides you with excellent visibility. You do not need to strike up a conversation every time he passes, but saying hello at least once a day is a good idea.

If you do not sit close to your supervisor's office, then getting social visibility will require more of an effort. For instance, consider the engineer who sits in a different building or is tucked away in a special laboratory far away. This person will probably seldom see the supervisor since it will take a lot of time and effort to do so. If this is your situation you must overcome the barriers caused by the separated location. Make a consistent effort to see your supervisor, at least once a week. When you meet with your supervisor make sure you are not just reporting technical progress— take time to socialize. Your supervisor is no different from anyone else. If you are out of sight, more than likely you are out of mind.

If your supervisor walks around checking on progress, make sure you are at least working and not socializing when the check comes. There is nothing like letting your supervisor catch you doing something important at your desk to help your image. This opportunity can occur more easily if you are sitting close. Also, often supervisors do not have time to hunt down people; they give assignments to the first available person they find. It's easy to find you when you sit just outside his door.

Sitting close to the office also provides other benefits. If you are within hearing distance you can listen to what is going on. If your superviser suddenly rushes out to the secretary announcing that he desperately needs some information and you just happen to have it, there is no better time to share it. Some supervisors are *not* very discrete about providing negative feedback. They loudly reprimand people in public for making a mistake. By sitting just outside your supervisor's door you can quickly learn what not to do. Correspondingly, you will also learn what pleases from the positive feedback given to other people.

Most of the good conversations in the company occur just outside the supervisor's office. You hear everything, from what the janitors are doing to what the vice president is doing. It is very informative to sit just outside your supervisor's office.

Sitting nearby also lets you see who is coming and going. People in the company are quick to inform supervisors about how poorly people are doing. This will give you the opportunity to find out which people in the company are going to report any bad news. It is always a good thing to know which people are constantly complaining and avoid them.

Another way to get good visibility is by where you sit in the conference room during a meeting. Sitting in the back row and acting like you really don't want to be there does not help your career. The optimal place to set is at the head of the table next to the person who called the meeting. Always sit next to the person who called the meeting if you can. This subconsciously gives people at the meeting the image that you are important. It also requires you to be alert and pay attention.

Social visibility also occurs at lunchtime. Where and with whom do you go to lunch? To illustrate the point, it is almost assured that you will progress up the ladder faster by going to lunch with the vice president. The point here is use your lunchtime to increase your social visibility in the company. Go out to lunch with your supervisor, the senior people, and technicians you work with. If people on your team get to know you and like you, the team will work better together.

Good social visibility is also obtained through your attitude. Your attitude tells people about you. Simply put, how do you view the world? Is the glass half full or half empty? The optimist claims the glass is half full. The pessimist claims the glass is half empty. Engineering is a difficult profession with many difficult problems to solve. It is very easy to turn into a pessimist and only see the problems. It is harder to be an optimist. Successful engineers are optimists, they find solutions to problems. Management rewards people who find solutions and make things work. Management does not give out awards to people who can only find problems with no solutions.

Your attitude will affect the way you treat people and, correspondingly, your social visibility. Some engineers act as though they are know-it-alls. Nobody else in the company has as much knowledge as they do. For these engineers, their social visibility is often negative and career limiting. Every person in the company likes to think they are important to its operation. Every person likes to think they have some unique knowledge. If you treat people as though they are important and have important knowledge, you will get along better with people. You will have good social visibility.

Part of what makes up your social visibility is your behavior at meetings. When you are in a group or team meeting look closely at your behavior. Are you allowing everyone to express their opinions? Are you quick to criticize and slow to compliment? Good social visibility at meetings requires that you follow good meeting etiquette. This means doing such things as letting people express their opinions, praising them in public, and critiquing them in private. If you called the meeting it is your responsibility to make sure you stick to the topic and do not get sidetracked. Poorly planned and poorly controlled meetings waste everyone's time. This is not good for the company and not good visibility for you. If you don't know how to run a meeting, then get some training.

Meetings are one way that senior engineers can see that junior engineers get some visibility. Often junior engineers are afraid to speak up and consequently contribute little to the meeting. By seeking out the opinions of junior engineers, senior engineers can help themselves and the junior engineer at the same time. Often just asking for the opinion of junior engineers is a boost to them. They find it stimulating that the senior engineer actually stopped to take the time to ask them for their opinion. And the senior engineer, much to his surprise, might find that the junior engineer actually does have valuable inputs.

Good social visibility also occurs in the lab. The engineer quickly learns there are things you can do and things you absolutely cannot do in the lab. For instance, never borrow or take lab equipment away from someone's experimental setup without asking them first. This socially inappropriate move will result in you getting your head handed to you and possibly barred from the lab. Taking credit for something that another person did in the lab is also bad visibility. Changing the experimental setup without letting anyone know can cause hours of lost work while others are trying to find out what has changed. Not following safe lab procedures simply because you are in a hurry may result in harm to other people. This is a socially bad move and can even get you fired. Not cleaning up after yourself is another socially bad move. Nobody likes to clean up other people's mess; don't make them clean up after you.

Use of the company computer E-mail network is another place for the engineer to improve his social visibility. How do you use E-mail, for company use or to socialize? Remember, you can say things in the hallway and nothing is recorded. However, if you write it down on E-mail there is a written copy. Do you take the time to spell-check your memos? Misspellings can send the wrong message about your work.

Social visibility also occurs through the way you meet and greet people inside and outside of the company. How sociable are you when greeting new employees and people from outside the company? It is always a good social practice to shake hands when you meet people for the first time. For people who are from outside of the company, you should exchange business cards. Business cards allow people to quickly see what your name is, how it is spelled, and your title. It provides them with a means of contacting you after the meeting is over. Business cards are not normally exchanged when meeting people from your own company.

Social visibility can be developed through the company's social functions. Is there an employee club that sponsors employee activities? Many companies have employee clubs that sponsor outings to baseball games, basketball games, football games, golf, theaters, and many more social functions. Supervisors and upper-level managers also attend these functions. There is no better way to meet these people than when you are relaxed, away from the office, and the pressure is off.

Socially visiting with upper management does not give you the immediate right to start talking about work. Often upper-level managers attend these functions to get to know their people socially and get away from the problems of work. Talking business during a social function could actually hurt you and result in bad visibility. The general rule is to talk business only if someone else brings it up.

Company-wide social visibility can be obtained by volunteering to work on community service committees. These include such services as blood banks, food share programs, United Way, and emergency medical response teams, just to name a few. Serving on these committees and helping to organize them goes along way toward improving your social visibility.

Often these committees are headed by upper-level managers, the same upper-level managers who do the promoting. You can help others out and at the same time provide your career a little boost. Check your company calendar for possible events that you might become involved with.

Finally, social visibility occurs at the national level during your attendance at symposiums and seminars. At these meetings you have the opportunity to socialize with colleagues in your field during social functions. Award banquets and social hours are very common. Everyone there is interested in meeting other people and making contacts. Being very sociable can help you discover other people in the field. You might use these contacts to find out about other job opportunities or the latest breakthroughs.

GUIDELINES TO ENSURING CORRECT USE OF VISIBILITY

Now that you understand the value of good visibility and the dangers of bad visibility, treat this tool or power with respect. Use it wisely and make sure you think before using it. It will take a considerable amount of good visibility to overcome one bad showing. I highly recommend you use the following guidelines to ensure that your visibility will benefit everyone.

Guidelines for Using Visibility

1. Use it to carry good news to upper management.
2. Don't use it to harm or blame other people.
3. Share the credit and visibility (include co-workers). Use "we" and not "I."
4. Clearly identify all who have contributed.
5. If you have visibility with upper-level management make sure the managers in between know that you are using it before you do it.
6. Good visibility is done in the open for all to see and not behind closed doors.
7. Think about the possible consequences before using your visibility.

If you follow these guidelines, your use of visibility should benefit not only you but everyone around you. You should not feel uncomfortable using it to benefit your career and others' careers. Remember, solid career advancement can only happen if you make sure everyone gets a piece of the visibility.

THE VALUE OF GETTING OTHER PEOPLE VISIBILITY

Getting other people visibility can be as beneficial to your career as getting visibility for yourself [16]. When someone does a good job for you do you just say thank you and walk away? If you do, you're missing a wonderful opportunity to get other people visibility. Why not ask the person for his or her supervisor's name and then send the supervisor a short memo or E-mail letting him know about the great job his employee did for you? Make sure you send the employee and your own boss a copy of the memo or E-mail.

Everybody likes to get a little recognition for their efforts. Believe me, this can also go a long way in helping your career. How do you think the person is going to react next time you go to them for help? More than likely you will probably get the best service you could image. People will be more likely to go out of their way, since they know that you appreciate their efforts and you are willing to make them look good in front of their supervisor.

Just imagine all the friends or relationships you will have developed over a long period of time. I'm sure your visibility in the company will significantly improve. The work you submit to another department will no longer be just one among many. It will more than likely always get special attention.

How about your lead engineer? Can you give her any good visibility? You sure can. The next time you meet with your supervisor is a perfect opportunity. I'm sure that your lead engineer at some time or another has put in a little extra effort to help you out. Isn't there something you can compliment her on in front of your boss? I'm sure you'll find something if you think about it.

How about your supervisor? Can you give him visibility also? You sure can. Catch him doing something you like and let him know about it. He is just like anyone else in the company, looking to be appreciated. Do you get the chance to talk to your supervisor's boss? There is no better time to pay your supervisor a compliment.

Giving visibility is like giving a smile. It costs you nothing to give. It helps everyone, and the more you give it away the more you get it back. So always give away some visibility and a smile whenever you can.

SUMMARY

The result of good visibility is accelerated career advancement. Bad visibility can seriously limit your career advancement.

Good Visibility = Power.

There is technical visibility and social visibility. Technical visibility is making management and co-workers aware of the engineer's skills and accomplishments for the purpose of benefiting the company, your career, and the careers of others. Social visibility is making management aware of the engineer's social skills and team leadership skills for the benefit of helping the company, your career, and the careers of others.

To be successful the engineer must have both technical visibility and social visibility. Obtaining visibility is something that an engineer needs to work on everyday. Technical and social visibility occur at the group level, company-wide level, and industry-wide or national level. *The best visibility occurs when everyone shares in the visibility.*

ASSIGNMENTS

1. Name one thing you could do this week to get technical visibility at your workplace. Name one thing that you could do for social visibility at your workplace this week.
2. Identify two ways you can become technically visible.
3. Identify two ways you can become socially visible.
4. Identify one action you could take to get someone else visibility in your group.
5. Give away some visibility.

CHAPTER 13

SUCCESSFULLY SURVIVING CORPORATE TAKEOVERS, MERGERS, SHUTDOWNS, AND REDUCTIONS IN WORK FORCE

Corporate takeovers, mergers, shutdowns, and reductions in work force are hazards for engineers. For successful career development you must know how to deal with and survive these career-damaging events. Your goal is not only to minimize the damage to your career but, in fact, walk away with a career advancement opportunity, if possible.

In this chapter we will explore how you can do just that. First we discuss what happens during takeovers, mergers, shutdowns, and work force reductions. Next we provide guidelines to assess how much you are at risk of losing your present job. Then guidance is provided on how to generate backup plans or make the decision to move on to better opportunities. Sometimes layoffs come without warnings. We show you what to do immediately should this happen and what to continue to do to successfully survive.

UNDERSTANDING WHAT HAPPENS DURING TAKEOVERS, MERGERS, SHUTDOWNS, AND WORK FORCE REDUCTIONS

Lack of knowledge about future events during takeovers, mergers, shutdowns and work force reductions creates additional anxiety and fear. This additional anxiety and fear can work against you. It can cloud your judgment and paralyze you, preventing you from taking action at times. *If you are aware of the sequence of events that normally occur during takeovers, mergers, and work force reductions this should help reduce the anxiety and fear, since you will know what is coming next.*

With a lower anxiety level and knowledge of the sequence of events, you are better prepared to plan your next move and take appropriate action ahead of time. I'd first like to summarize some of the common events that occur in companies during a takeover, merger, shutdown, or work force reduction. Let's first discuss what happens during work force reductions and then explore takeovers, mergers, and shutdowns.

The first indication that trouble is on the horizon is the profit line and stock value. If the profits are falling and your company's stock price has suddenly fallen these are the first indicators of trouble ahead. If the profits continue to fall companies will not be able to stay in business. Management has to take some type of action to turn the trend around. In response to profit and stock value problems most companies will announce a series of predictable actions to be carried out. These actions are usually identified as cost cutting measures and reorganization.

The cost cutting measures will include phasing out old product lines that are no longer profitable and reorganizing to help reduce costs and streamline the company. Often the corporate staff is cut or pushed down the chain so it does not appear that the company has so many vice presidents. A hiring freeze might be announced. The company might reorganize to eliminate any duplication of effort. For example, having one stockroom instead of two, one print room instead of two, one lab instead of two, and so on. These are all good for the company.

If profits still continue to fall, then it is on to the next level. Reduction of the work force through early retirement incentives and elimination of part-time and temporary help where possible. In addition, several departments may get combined and duplicated jobs eliminated. Usually at this stage there are some layoffs of the poorest performers or people who have absolutely no work in their department.

If the profits still continue to fall or the company's markets decline, then it is on to the next level of trimming. The work force is rated and totem rankings are made. Those on the bottom of the totems become the candidates for the next round of work force reductions. Managers will often protect those who have been loyal to them. This may not result in saving the best people, only the favorites. Upper management is aware of this practice and usually takes steps to counteract it. To counteract this upper management will reorganize the departments and assign new supervisors who are not interested in loyalty but only in doing whatever it takes to turn the company around. If this means laying people off, then so be it. These hatchet men, as they are sometimes referred to, quickly cut the unwanted help with little attention to what has happened in the past. It is easier for them to do this, since they have no personal ties with the employees they must supervise.

If your supervisor should be suddenly replaced during one of the reorganizations, you have your work cut out for you. It is almost like starting a new job. You must start all over to prove yourself to the new supervisor. This will take time and effort, so start immediately. If you recognize this early and start immediately you will be able to prove your worth to the new supervisor in a shorter time, and, hopefully, keep your name off the layoff list. Ignoring the fact that your supervisor has changed and assuming that your past record will stand is not going to help keep your position. When your supervisor changes it is like the clock has been reset and everyone starts over.

The best thing you can do is react as though you just started a new job. You need to interview. Take time to sit down, one-on-one. Discuss what your accomplishments were in the past and the value you bring to the group. Provide a copy of your résumé and show your portfolio. Do not assume that your supervisor will automatically do research and review your personal file. As one mentor of mine once said, "Never assume! It makes an ASS of U and ME." Unfortunately, I still had to discover this the hard way on several occasions.

If the profits continue to fall, this process of reorganizing and cutting continues until the company either becomes profitable or it folds. The company morale will cycle up and down with each wave of layoffs. Each time there is a layoff announcement the morale usually dips to another low. Everyone is worried and the efficiency of the organization can at times come to a grinding halt. Workers are talking about who will be the next to get laid off and no one seems interested in their work. The best thing you can do is keep your efficiency up and do not participate in the hallway discussions about who will be the next to go. Your time is better spent on your work and your backup plans.

Soon after the layoffs have been completed and those laid off depart, the morale will start to climb again. The hopes and expectations are that the profits will increase and the layoffs will stop. Not to cycle up and down with each wave of layoffs and reorganization is impossible. These are very tough conditions to work under. If you are going through this and feel you cannot handle it, then get professional help. Often companies will have psychologists on hand to help people deal with the stress. If your company has a psychologist available, seek out this help if you need it. Sometimes just being able to talk to someone can be a tremendous benefit. They are trained professionals who can help you. Use this company benefit if it is available to you.

During company mergers several different events may occur. The first step toward a merger is the preparation of a statement of net technical worth by the company executives. This is simply a sales brochure describing all the valuable assets and contracts the company possesses. Next comes the big announcement that company officials have decided to merge or sell controlling interests to other companies to remain profitable. At this point the company is literally up for sale. Those companies interested in purchasing the company now take tours of your facilities to see for themselves the assets of the company. The tour groups will include bankers, lawyers, executives, and, in some cases, engineers from other companies to help assess the company's technical worth.

After the tours comes the big announcement: the company has been sold and there will be a merger if the Federal Trade Commission approves the sale. During this time many people from the purchasing company will tour your facilities and interview selected people to determine what stays and what gets axed. Then the big day arrives and the sale has been approved by the FTC. An official takeover date is announced.

From this point on, generally one of four things can occur. The first and least damaging is that the name on the building is changed, a few people are let go, and the operation stays pretty much intact. The second is that major reorganizations will occur. There is a major influx of people from the new parent company to teach everyone how things will be done in the future and there are a large number of layoffs. Those people getting laid off are usually the ones the parent company determined through interviews are not needed. They may not be needed because the parent company already has people doing their jobs or it does not want the type of business that would utilize their skills.

If you are selected for an interview make sure the interview goes both ways. Not only tell the new parent company about your skills but make sure you find out if they need your skills or already have a department doing the same operations. This will give you insight as to whether your department and you will survive the merger.

The third thing that can occur is that the company is dissected. This means the company's operations will be cut up and separated. The new parent company usually will transfer parts or operations of the acquired company to other locations. Operations may be transferred to other states where they will be used to support or strengthen existing operations in the new parent company. If this occurs you may be asked to move to the new facility if you wish to keep your job. Operations not needed are not transferred and the people simply let go.

The fourth and final thing that may occur is complete shutdown. In this case the new parent company lays everyone off and closes the doors. This occurs usually for two main reasons. One reason is that the assets of the company are more valuable than the engineering work. For instance, the land may be more valuable than the building or the products it produces. The new company can make a quick profit by shutting everything down, letting everyone go, and selling the land. Another reason is that a competitor has bought out the company to get its contracts. The new owners can handle the contracts with its existing facilities and do not need the acquired facilities or people. Therefore they simply transfer the contracts and shut down the company, letting everyone go.

The steps I have described that occur during takeovers, mergers, and work force reductions have been generalized and simplified. The actual events that occur will vary from company to company and situation to situation. Usually it follows the sequence that has been described. Hopefully, these descriptions are enough to make you aware, allowing you to make better decisions about your future.

Now that you understand the steps that occur, let's look at some guidelines you can use to determine if you are at risk of losing your job.

DETERMINING IF YOU ARE AT RISK OF LOSING YOUR JOB

The first step in successfully surviving is to determine if you are at risk of losing your present job. Once you have determined the risk involved then you can develop a plan and react appropriately to ensure your continued employment. To help you in determining how much you may be at risk in your present job situation, I have come up with the following list of questions. Please keep count of the number of times you respond with a "yes" answer as you read the questions.

1. Does your company have high debt and a low cash flow?
2. Has management already tried an early retirement work force reduction effort?

3. Is there a hiring freeze on at your division?

4. Has your company merged with another that duplicates your work?

5. Has your supervisor or his supervisor been laid off?

6. Is the health of your industry poor? Are people ordering less of your company's products?

7. Has your management announced cost cutting measures to be implemented?

8. Is your supervisor constantly revising the department workload forecast for upper management to review?

9. Have there already been layoffs at your plant or division?

10. Are you on the lower portion of the employee totem?

11. Is your supervisor or program manager forecasting an end to a contract with no replacement?

12. Are more layoffs forecasted for the division?

13. Did you receive a below average rating on your last employee appraisal?

14. Has your supervisor announced there will be a layoff in your group?

15. Has the marketing department stopped advertising your product?

16. Is there a plan to phase out the product line you are working on?

17. Has the company announced a loss for the last quarter?

18. Is your supervisor meeting with the personnel department on a steady basis?

19. Is your supervisor's door always shut when meeting with the personnel manager?

20. Are other departments doing the work that you normally do?

21. Are you not getting invited to meetings that you normally used to attend?

22. Has the building maintenance department stopped working on your area?

23. Is your supervisor forecasting a work force reduction?

24. Is your equipment being transferred to other groups or divisions?

25. Has your supervisor's secretary suddenly stopped talking about the work force reduction to you or anyone else?

26. Are there several people in your group doing exactly the same work?

27. Are there several people in your group at exactly the same level?

28. Are you the highest paid senior person in your group and can lower-level employees perform the same work?

29. Are you the most junior person in the group with the least experience?

30. Do you have poor relations with your supervisor?

31. Are you doing "make work" assignments that do not really contribute to the company's profit line? (For example, writing procedures, manuals, or standards.)

32. Was your company bought out by another one in the last six months?

33. Have you lost interest in your job?

If you have answered yes more than 20 times then you probably are at great risk of losing your job. The chances are excellent that you may be part of the next work force reduction. You should be on what the military calls red alert. If you have answered yes 10 to 20 times you should be on yellow alert. There is a fair-to-good chance that you may be next in line for layoff. Both red and yellow alerts indicate that you should be taking backup steps to ensure your employment in the future. If you answered yes less than 10 times, then the chances are you are not in immediate danger of being laid off, but you should still continue to monitor the situation.

What should you be doing if you are on red or yellow alert? Get busy! The first thing you need to do is some investigating to determine how much you are at risk, and there is no better place to start than your supervisor's office. Simply reserve some time on his or her schedule and have a heart-to-heart talk with him. Start out by clearly identifying that you are concerned about your future with the company. If she stops you right there and lets you know that you are not in trouble, that is a good sign. Get him to expand on why he thinks you are not in trouble. Has she been told her department will not be affected by cutbacks? Has he reviewed the workload and feels there is enough work? Has she spoken with her supervisor about the layoffs and who they will affect?

If he leads you to believe that work force reductions are coming, ask him straightforwardly, "am I on the layoff list?" He will have one of two answers: "No" or "I am not allowed to share that information with you." Each answer tells you what you need to know. The first answer indicates you are not in danger for now, the second answer says your name is being considered.

Probe further; do not leave without a clear understanding of exactly how he feels things are going. Even if your supervisor leads you to believe that no layoffs are coming you cannot simply take his word for it. Sometimes supervisors themselves do not know simply because they will also be part of the work force reduction. Upper management does not bother to tell them, since they need the supervisors to keep functioning until the end comes. In certain cases supervisors may fear that if they say anything it could throw the entire group into chaos. They quickly conclude that it's best to be silent until the end comes. In any case, once you leave the office you still have further investigating to do.

Utilize your other contacts in the company to find out what might be going on. Check with your mentors. Do they know anything that they can share with you? Visit with the group accountant and pick up a copy of the workload forecast for your group. Is she forecasting fewer employees in the months ahead? What are the grade levels that will be cut? How about the last totem taken? Is there any chance of finding out where you stood on it? If you were on the bottom for your grade level, then it is highly likely that you are being considered for layoff. Again, visit the program manager on your project. Does he consider you critical to the program? If you have been identified as critical to the success of a project, you are usually not laid off.

Tap your social connections in the company for any information they have. People in other groups often hear that it's not their group this time but it will be a certain other group. Can you tap any other social contacts you made in the company, perhaps people on the committees you served on? Often they will share things with you. In any case, if you hear something check the source of it and do not take it for granted. It is unbelievable how rumors get started and circulate around the company.

One upper-level manager started the latest-rumor sign-up sheet outside his office. He simply put up a blank sheet of paper and told the employees to write down the latest rumor they heard and he would tell them if it was true. He did this during a merger that was going on with another division. Within days he had over thirty rumors that ranged the spectrum from everyone getting fired tomorrow to raises for everyone since the merger was going well. The truth of the matter was that nearly every rumor was just that, a rumor not based on any fact. For something as serious as your employment make sure you double check and get verification from multiple sources before you do anything.

The ultimate objective of all this research is to come to a conclusion about your future employment with the company. To help you better understand where you stand, gather all the information together and write it down. Diagram what you know; list the positives and list the negatives. If the negatives clearly outnumber the positives this will tell you something and aid you with your decision. The conclusion must be based on facts and then you must decide to either stay and take your chances or start looking for a new job. Only you can make this decision.

If making a decision like this frightens you, you can take some comfort in knowing that you are not alone. Everyone is frightened when it comes to making a decision as big as this. Most of us do not want to leave a company that we have worked years for or to change jobs and start all over. The thought of it is very frightening. However, you must decide to do one or the other. If you ignore the situation, you are setting yourself up for worse failure by letting others control your career. By choosing one course or the other you are taking control of your career and you are determining your destiny. You will fare significantly better if you are in control. Let us now look at some actions you should take when you choose one versus the other.

DECIDING TO STAY AND MINIMIZING YOUR CHANCES FOR LAYOFF

If your decision is to stay with your present employer since you feel the risk is small for your being laid off, you still cannot relax and lay back. You must work toward minimizing your chances for being laid off in the future. There are actions you can take to minimize your chances for being laid off [17,18]. Taking these actions will not automatically guarantee that you will be immune from future layoffs. They should, however, help you reduce the chances.

The first action is to critically evaluate your recent performance. Are you meeting deadlines? Has your work been above average? Are you getting good visibility with your supervisor and the managers? If you are not getting good visibility then you should change your work habits. If it requires you put in extra effort, commit yourself and just do it. It doesn't hurt to put in extra effort if it results in job security. Showing up for work early and leaving late will also help. Volunteering for extra assignments so that you become critical to the company is another action.

During work force reduction everyone is on edge due to the uncertainty. As a result people are nervous and tend to easily end up in arguments. Don't start any shouting matches. The last thing you need is your co-worker or supervisor thinking how difficult it is to work with you and how you will not listen to reason.

Make sure your work is the best it can be. Remember, it will be compared with others in your group. Spending a few extra minutes making sure your work is neat, organized, and clearly communicated is absolutely essential. Do not leave anything half done—carry it to completion. Before you present it to your supervisor, try to think of questions or criticism that he or she may have, based on other work you have done. Anticipate these things and have answers ready. Don't present problems without solutions. Supervisors are looking for engineers who are willing to work on the problems, not complain about them.

Make sure you get credit for all your work. Often engineers will help other engineers at the expense of not getting their own work done. This is not a bad thing to do, but I recommend that you don't help others out unless your supervisor knows about it. To help others at the expense of not getting your own work done is career limiting to start with. Doing this during a merger or work force reduction can be fatal to your career. I'm not saying not to help other engineers out! Everyone needs help from time to time. What I am saying is, if you are going to put in the extra effort to help someone, make sure you also put in the extra effort to tell your supervisor. During work force reductions your supervisor must decide who to keep and who to let go. He will make this decision based on your visible and demonstrated performance. It is essential to make sure all your work is visible to him.

If your group is going through a work force reduction but other parts of the company are strong, then try to transfer to another group. Get busy reading the company job ads. Maybe there is another group that can use your talents. To help you find out about other jobs, talk to your contacts, your mentors, and any other persons who might know about openings in other divisions. Normally the personnel department is doing this before you even learn about it. Most companies try to transfer people to other groups rather than lay them off. Often you will be notified by your supervisor that you are being transferred to another group. Cooperate and make the move if it is a good one. But make sure you check into everything that is involved before you commit to moving, as discussed in Chapter 10, on career moves.

MAKING BACKUP PLANS JUST IN CASE

Even if you decide to stay, you should develop some backup plans just in case things quickly take a turn for the worst. [19]. This should not be foreign to you since all good engineers make backup plans in case their work encounters unexpected problems. This is commonly called contingency planning.

The first thing your contingency plan should include is updating your résumé. Go to the library and get several books on how to write your résumé. Start reading and start updating your résumé.

Second, start looking through the job ads outside of your company. You do not need to respond to them but clip and save them in a folder for future reference. If you are considering making a move to a new location, contact the major newspapers in that area and order the Sunday paper with all the job listings. Clip the job ads from the trade journals.

Third, identify any co-workers or previous supervisors in your company you might get a letter of recommendation from. These may be people with whom you worked in the past who have complimented your work and might be willing to write a good recommendation letter. You need not contact them but at least make a list. In the future, if you suddenly need a letter of recommendation you will know who to talk to. If you feel comfortable enough about contacting them, you might ask, "In the future, if I ever need a letter of recommendation would you write one for me?" If they respond with a yes, you're all set for now. If they respond with a no, then you know who to stay away from.

Fourth, start a portfolio of your work and awards. Put into a binder any photographs of significant projects you have worked on. Awards and certificates are good to include. You can think of this as a scrapbook of your accomplishments. My personal experience has shown this to be one of the most valuable interviewing tools. Most people I've shown my portfolio to have told me that I was the *best prepared candidate they ever interviewed*! The portfolio clearly showed all my experience, something that I found extremely difficult to do with words. To quote a very old saying, "One picture is worth a thousand words." To this I would like to add, "Several good pictures in a portfolio gets a much better job offer."

> Being well prepared is the best contingency planning you can do!

If you make the decision to leave, since you feel the risk is too great in staying, then you must start immediately preparing yourself to leave. This will not be easy, since you will be constantly fighting back the temptation to stay. I found myself minimizing the risk and telling myself, "They would never lay me off; I'm too important to the project." "I have 12 years in with the company, I do not want to lose my benefits, especially my three weeks vacation." I quickly found out how wrong I was when the company downsized by 50% and I was part of it. Making the decision to leave will be frightening, but let's look at how you can make the transition easier.

WHAT TO DO IF YOU DECIDE TO LEAVE THE COMPANY

The first steps you should follow once you have decided to leave are very similar to the ones recommended for backup plans. First, start your portfolio. Gather all the photographs, impressive diagrams, and awards that you have. Organize them into an impressive binder that tells a story of your work and your accomplishments. Next, search out people who will provide you with good letters of recommendation. Approach these people if you can and get them to write a letter. A suggestion I have is to write down what you feel would be important for them to say in the letter. Give them a copy of the list and ask if they might highlight the things you have come up with, since you will be stressing those skills in your résumé. It's always good to give them the freedom to add others if they may think of any.

The next action is to start rewriting your résumé. For this, go to the library and pick up several books on how to write résumés. Spend some time studying the different approaches and pick at least one or two styles that you will feel comfortable with. Before you actually start rewriting your résumé, you must spend some time thinking about your future career and the direction in which you wish to go. Maybe it is time that your career took another direction and you changed your area of interest. Once you have made a decision about the direction you are ready to start writing your résumé. Once you have completed your résumé it is now on to job interviewing. Refer to the next section where we discuss the job search.

Before we move on, I would like to share two very interesting observations with you. The first has to do with your ability to find a new job. It is a well known fact that it is easier to find a job if you already have a job. This would indicate that if you feel that you are going to be laid off, the best chances for finding a new job would be while you are still working. So if you find yourself in this position, take immediate action!

The second observation has to do with the timing of your departure. If you leave before you are laid off you give up any severance package that you are entitled to. These severance packages can be very significant. For instance, some companies offer up to two weeks severance pay for each year of service. If you have been at a company for 15 years, this means you are entitled to six months of pay when you leave. Based on this, the optimum time to interview is before you are laid off and the optimum time to start the new position is within a few days after you have been laid off. If you can time everything just right you can walk away from one company, receive a bonus and start at the new company with a raise. I have personally seen this happen.

One engineer I know started interviewing before he was laid off and found a new job. He timed it so that he started the new job just two weeks after he was laid off. Since he was laid off, he received a severance package of several months of pay. With the new position he received a slight pay increase. To summarize, that year he received an extra paid two-week vacation and banked a two-month severance pay package as a bonus for going out and finding a new job. Not only that, the new job was closer to his home and less of a commute for him. Now that is really surviving a layoff! If you have trouble getting started once you have made the decision to leave, simply think of it as getting yourself a well deserved bonus. This should help you keep going!

WHAT TO DO IF YOU ARE LAID OFF

If you are laid off the question becomes what should you do. Before we can discuss what your best actions will be, you must first understand what you are going through. When you are laid off, fired, or let go, work force reductions and downsizing become personal and overwhelming. Normally, no one is prepared for the shock. Nothing like this had ever happened to me before. I was totally numb and felt paralyzed with shock. Luckily a company counselor was at hand and she sat with us the first few hours after we were notified to help us get through it.

Losing your job will result in a grief reaction. This reaction will be similar to the grief reaction that a person experiences when loosing a loved one. Your loss maybe the loss of self-worth, the loss of co-workers, the loss of security, or the loss of responsibility. You will be experiencing some or all of the following emotions:

Anger	Guilt
Depression	Fear
Bargaining	Denial

There is no set order in which you will experience these emotions. You may move from one emotion to another with no set pattern. The goal is to recognize them, accept the situation, and move on to more constructive activities as soon as possible. Even after you have reached acceptance, you will continue to reexperience these emotions from time to time.

The important thing to recognize is that you will be under great stress and you need to take action to effectively deal with the stress if you are to successfully survive. Some effective ways in dealing with the stress include:

Reestablish a routine in your life (such as the time you wake up and go to bed)

Develop and utilize a support system (family, friends, counselors)

Exercise regularly to relieve stress

Eat regular, balanced meals

Reward yourself with enjoyable activities

Get plenty of rest

Talk to others; share your feelings with family, friends, and counselors

Similarly, there are things you should not be doing that will hurt your progress and ability to accept reality and move on. These include:

Excessive use of alcohol

Blaming people (your co-workers or family)

Running away from your situation

Working on something to the point of exhaustion

You now need to stay alert and deal with problems as they occur in order for you to survive. More than likely there will be other stresses occurring in your life at the same time. These may include death of a loved one, divorce or separation, injury, marriage or financial problems, children leaving home, and change in residence just to name a few. Dealing with a layoff is extremely tough; dealing with a layoff as well as other stress can be devastating if you lose control and never reach acceptance. Your goal is to reach acceptance and move on to constructive actions that will lead to a new job opportunity [20].

If you are laid off you need to keep a level head and immediately take action. Normally, when you are laid off you are given a termination date that may range anywhere from one day to two weeks or more. Immediately upon learning of your termination, you need to start taking the following actions before you leave the company for good. These actions will be hard to do, but to successfully survive, you must. They usually involve your supervisor or personnel department.

First make sure you will get your vacation pay. You are entitled to it, you have earned it, you deserve it! [21,22]. Next find out if there is a severance package. Most companies offer a severance package based on years of service. If nothing is offered, then start bargaining. You should be able to shame them into giving you something. You will need this money to support yourself until you have found a new job.

Next, you need to make sure your vested rights in profit sharing and/or pension still stand and the money you have accumulated is not affected by the layoff. Contact your personnel department immediately and discuss with them how to handle your account. Do not bypass this; it could end up costing you money if you mistakenly withdraw from your IRA or other plans.

Get an extension on your medical insurance. Most companies will offer you the opportunity to continue your medical and dental benefits, but you will have to pay for it personally. Take the coverage; it is usually the best coverage available. If the company does not offer an extension on your benefits, then contact your health insurance carrier directly. They often have plans available for people who have lost their jobs due to layoffs, but, again, you will have to pay for it personally. The monthly premiums may seem expensive compared to what you were paying, since you are without an income. But with no insurance coverage you will be wiped out if a major medical emergency occurs to you or one of your family members.

You will be experiencing an overwhelming amount of self-doubt, pain, and anger. Put those feelings aside; you still have more work to do before you leave. Because of this self-doubt and anger you will not want to be around your co-workers or even talk with them. The best thing you can do, however, is just the opposite. Tell everyone of your situation. Now is the time to utilize your contacts—every contact you have made over the years, if possible. Leave no stone unturned, as they say. Contact everyone and anyone in the company who may have a job lead for you. Do not let your ego get in the way and become a roadblock to an even more successful job. Don't be afraid to ask for names; it is surprising how people will open up and help you once they are aware of your situation.

You may find that some people will help you simply to bolster their own egos. It provides them with an opportunity to show off who they know and how important they are in the industry. Others may help you because they have been in the same situation themselves and know how hard it is. Whatever the case, let them help you. If it gets you a job, isn't it worth it?

When someone gives you a lead make sure you get all the information you can about the opportunity. Who is the person doing the hiring? What kind of help are they looking for? What is the salary? You need to ask these questions for two reasons. First, you have to determine if this is a good lead for you. Often people will provide you with leads that are not what you are really interested in. If you blindly start chasing every lead without finding out how good they are, you will be wasting precious time. Second, if it is a good lead you will want to go to the interview as well prepared as you can. You will know in advance what they are looking for. This allows you to emphasize the skills you have that match the job requirements. This is always a plus in your favor during an interview.

When someone provides you with a lead, always check with the person giving you the lead to see if it is all right to mention their name. Their answer to this question will help you qualify the lead. If they only heard about it through the grapevine and have no direct contact with the person doing the hiring, then you know this is a long shot. However, if they personally know and are good friends with the person doing the hiring, then you know this is a good lead to follow up on immediately.

Who should be the first person you start asking for leads? The same person who just laid you off. He or she may even offer to furnish you with some job leads; if not, ask. From your supervisor it is on to your close co-workers, mentors, and any other contacts you have within the company. You do not know where the lead that lands you a new job will come from. Therefore you have to contact everyone.

As you search around the company, be sure to ask key people, including your former supervisor, for letters of recommendation. You should ask several people for letters; not everyone you ask will get around to writing one. In any case do not walk out of the company without at least three letters of recommendation. Sit in their office and wait for them to write and type it if you have to. You need to be persistent because the chances of getting a letter of recommendation decrease significantly once you walk out the door.

If you remain calm and follow these steps you will be leaving the company with several hot job leads and several letters of recommendation. This is very important! The leads will provide you with hope that other opportunities do exist and provide something for you to look forward to. If you are lucky the job leads could result in a new job opportunity, but do not pin all your hopes on them. The letters of recommendation will help bolster your sagging ego when you read how much people thought of your work. Between the two you have established a foundation from which to start your new job search.

Don't burn the bridges when you leave. Leave on a positive and upbeat note. If you feel up to it, visit with old co-workers and friends to say goodbye one last time. Spending some time saying goodbye is good for you. With everyone wishing you good luck, it helps you to start moving on. It also prepares you mentally for the changes that are to occur. Often people have the hardest time dealing with the loss of a loved one when they did not have the opportunity to speak with them one last time. Take the time to say goodbye. You should also remember these may be the same people your future employers will be contacting. It is of no value to leave on a bad note.

Cleaning out your desk will be very hard to do. Before you leave make sure you take any material that could be of use to you at a future job. If you are unsure as to what you can take, check with your supervisor. Do not take any proprietary information with you; this is industrial espionage. It is punishable by fines and prison terms.

Where should you be headed when you walk out the door? To either the company-sponsored job placement center, an employment agency, or someplace where you can work on your résumé and make phone calls on the leads you have. You have now successfully made the transition to the job search phase.

CONDUCTING A SUCCESSFUL JOB SEARCH

The first step in conducting a successful job search is to start with a self-inventory [23,24]. Simply put, you need to make a personal assessment of your job skills and interests. It is very important for you to know what your strengths and weaknesses are. Make an inventory of your abilities, interests, and significant past accomplishments. This inventory will make the job search easier. To help with the inventory process refer to Figures 13-1 through 13-3.

Shown in Figure 13-1 are three different lists of abilities or skills that engineers utilize on the job. These abilities or skills have been divided into technical, management, and personal skills. You should first review the lists and mark the skills that you feel are your strengths and those that are your weaknesses. Next identify, if you can, a specific past accomplishment that clearly demonstrates that you have that skill.

Type of skill	Strength or weakness	Past accomplishment demonstrating strength
Technical		
Hardware design		
Hardware build		
Laboratory test		
Laboratory research		
Technical publications		
Computer modeling		
CAD design and modeling		
Analysis and modeling		
Experimental research		
Patents		
Technical awards		
Programming		
Producibility		
Manufacturing		
Project Management		
Planning		
Budgeting		
Organizing		
Developing policies		
Developing procedures		
Controlling		
Cost tracking		
Schedule planning		
Customer interface		
Team formation		
Salary administration		
Department budgeting		
Capital planning		
Presentation skills		
Interpersonal Skills		
Motivating		
Team leadership		
Conflict resolution		
Work relationships		
Meeting skills		
Versatility		
Team dynamics		
Communication style		
Customer relationships		
Social abilities		
Mentoring		

FIGURE 13-1 Job abilities and skills inventory.

When you are done, you will have identified those skills that you consider your strengths and past accomplishments that you can point to during an interview that clearly demonstrate that you have this skill. This list becomes the starting point from which you will generate your résumé.

The next step is to try to visualize the type of job you wish to move into. To help with this visualization complete the interest inventory given in Figure 13-2. This inventory will help you identify those things that are important to you in a job. It will also help you to identify those things you wish to avoid in a job. Sometimes, when you are unsure of what you would like in a job, just listing the things you don't want in a job will help.

Visualization of your ideal job

1. Job title: _____
2. Job salary: _____
3. Company location: _____
4. Home location: _____
5. Commute time: _____
6. Company size: _____
7. Company products: _____
8. Size of engineering group: _____
9. Job functions: _____
   _____
   _____
10. Office size: _____
11. Company benefits: _____
12. Freedom to work on: _____
13. Travel: _____
14. Laboratories look like: _____
15. Personal computer: _____
16. Supervisor who: _____
17. Co-workers who: _____
18. Career advancement paths leading to: _____
   _____
19. In 5 years I'll be doing: _____
   _____
   _____

Visualization of your worst job

Things I absolutely will not put up with on my next job: _____
_____
_____
_____

FIGURE 13-2   Interest inventory.

You will use the interest inventory to help you sort through the job ads. By looking at your interest inventory and closely reading the job ads you should be quickly able to eliminate those jobs you are not interested in.

After you have completed the interest inventory it is on to the significant accomplishment inventory. This inventory sheet is shown in Figure 13-3. It is simply a listing of your most significant accomplishments and the skills that were demonstrated by the accomplishments. This inventory is very important. This information is what you should be putting on your résumé and discussing during interviews.

Once you have completed your inventories, it is time to start writing your résumé and looking through the job ads. The optimum way to do this is to customize your résumé for every job you apply for. Writing a generic résumé is not the best approach. Closely reading the job ad and tailoring your résumé to the skills called for is the best approach.

There are a million and one approaches to writing résumés that get results. The best bet is to go to your local library and pick up a few books on the subject. Read several different books, identify a style that you feel comfortable with, and use it.

My significant accomplishments were:	Abilities and skills demonstrated
1. _____	1. _____
2. _____	2. _____
3. _____	3. _____
4. _____	4. _____
5. _____	5. _____
6. _____	6. _____
7. _____	7. _____

FIGURE 13-3 Inventory of significant accomplishments.

With your inventories completed and a good understanding of how to write a résumé, it is time to start searching for job leads. Where do the leads come from? Everywhere! Get busy and start looking. Here is a list of places to start looking for job opportunities.

1. Contact all your business associates outside the company. This should include people in technical societies, vendors, or subcontractors you have worked with.

2. Contact every headhunter or employment agency that will take your résumé and work with you. Those employment agencies that specialize in your area are the best. You will usually find these agencies advertised in technical trade journals. Be careful; sometimes this can backfire for several different reasons. One example is when a company gets your résumé from several different headhunters. This can complicate the situation. The second is that if other headhunters find out that someone else is pushing your résumé around they may drop you.

3. Go to company-sponsored job placement centers. These centers are excellent for picking up leads. They often have a large number of free trade journals and newspapers for you to review. In addition, other people using the center will often share information about who is hiring and who is not. Some job placement centers even provide staff to help you with résumé writing, typing, mailing, and free use of phones. Make sure you take full advantage of this service if it is available to you.

4. Run your own ad in the trade journals. This can get your name in front of thousands of people.

5. Contact your college alumni association.

6. Attend trade shows and post your name.

7. Attend job fairs. Go prepared to discuss your qualifications and bring a complete résumé package.

8. Contact the major newspapers in the cities you are interested in working in and get a subscription to the Sunday paper.

9. Tell your neighbors and friends you are looking for a job.

10. Contact any professional societies that you belong to.

11. Contact Federal and State employment agencies.

These are just a few of the avenues available to you to find job opportunities. I'm sure that when you put your mind to it you can think of at least a half dozen more ways to generate job leads. I'm also confident that if you really put the effort into it, you will quickly find there are more job opportunities than you thought. In fact, there will be more than you can possibly handle. You will probably have to sort through the listings and leads and prioritize which ones you consider the best. If you get to this stage you are doing all the right things.

The details on how to write a résumé and how best to prepare for an interview are beyond the scope of this book. The reader is directed to your local library for material covering these subjects.

PROTECTING YOURSELF WHILE JOB SEARCHING

The job search phase can last anywhere from a month to over a year. During this time you must take steps to protect yourself [25]. The first step is to establish a new financial budget. Determine the minimum amount you can live on and start cutting your expenses immediately. Determine how long you can survive before you will deplete your savings. This will give you a deadline by which you must have a new job. Sometimes a deadline is an excellent motivator.

Lock up your credit cards immediately and start to make minimum payments wherever possible. Eliminate or delay any payments you can. Sometimes you can write a company in advance and notify them that you cannot make this month's payment but you plan to make it next month; it will not affect your credit history. Conserve energy whenever possible. Delay purchases on clothing and anything that is not absolutely essential. You may even consider selling a few things if you have to.

After being gainfully employed you are entitled to unemployment benefits. Go down immediately to the unemployment office and register for your benefits. It may take several weeks before you receive anything, so do it immediately.

Is there any part-time consulting or part-time jobs that you might be able to obtain? This will provide you with some income and still give you time to do some job searching.

A final note on discussing the bad news with friends and family members. The best thing you can do is tell them immediately. You may be able to hide it for a while but not forever. You will need your friends and family for support in the days ahead. Give them the opportunity to help you. Let them know exactly what happened, explain that it may be a while before you get another job, so you are counting on their support and cooperation.

You cannot effectively look for another job and hide what you are doing at the same time. Other companies will be calling you at home, so it will only be a matter of time before the family finds out.

SUMMARY

Successfully surviving takeovers, mergers and work force reductions requires you to be alert and actively assessing the situation. You must first determine if you are in danger of being laid off. Access the warning signs. If there is danger, at what stage are events at and how much of a danger do they pose to you?

If you are in danger you must make the decision to stay or move on. If you decide to stay, then you must take action immediately to secure your position as well as develop backup plans, just in case. If you decide to leave then you need to get busy looking for new job opportunities.

If you are laid off you will experience the grief reaction. This reaction will be similar to the experience of losing a loved one. You will be experiencing anger, guilt, depression, denial, and fear. To successfully survive you must recognize these feelings, deal with them, and move on to acceptance and more constructive activities. Before you leave the company, you must go into a high energy state of making sure you get all your benefits, any letters of recommendations you can, and a list of potential job leads.

Once you leave, you must continue in your high energy state. You need to take an inventory of your skills and interests. After completing these inventories it is on to résumé writing, job searching, and interviewing. To successfully survive you will need to make use of all your resources. These include friends, family members, professional contacts, and employment agencies, to name a few. You will not know where your next job will be coming from. Therefore you need to leave no stone unturned in your job search. If you are doing things right, your job search will, hopefully, require you to sort and prioritize all your opportunities.

ASSIGNMENTS

1. What are some of the warning signs that your job may be in danger?
2. What is the sequence of events during a takeover?
3. When is it better to stay than to leave? When is it better to leave than stay?

4. What is the grief reaction?

5. What makes a good letter of recommendation?

6. Does your supervisor always know what is going on?

7. Name two things you must do before leaving the company for the last time.

8. Complete the inventory sheets.

9. Update your résumé.

CHAPTER 14

GETTING ON THE FAST TRACK FOR ADVANCEMENT

If you want to move ahead quickly you have to become a member the fast-track crowd. The fast-track crowd is acutely aware that working long hours is simply not enough [26]. It is just the minimum requirement for keeping a job and not the means to fast-track advancement. The avenues to fast-track advancement lie in another direction, a direction that requires skills in selecting your supervisor, your assignments, and progress reporting. Let's identify some fast-track actions.

SELECT YOUR SUPERVISOR

The quickest way to advance is to attach yourself to a fast-rising supervisor and ride along. Hopefully, every time your supervisor is promoted, you are also promoted. For this to happen you must select and work for a supervisor who is clearly on the way up.

Fast-track supervisors do not wear name tags identifying themselves nor do they come along that often. You have to be able to recognize them and seek them out. You can identify fast-track supervisors by their great visibility with upper management. You can also identify them by the fact that their groups usually receive most of the special assignments from the vice presidents. Another characteristic trait is that their groups usually receive more awards and recognition than most groups.

Once you have identified a fast-track supervisor you need to get yourself transferred to his or her group any way you can. Once in the group you stand a better than average chance for fast-track advancement. Fast-track supervisors generally tend to rate their group members very high. This gives upper management the impression that the group has superior performance—*superior performance that is a result of the fast-track supervisor's outstanding efforts.* The fast-track supervisor realizes the value of overrating his or her people rather than underrating them.

Another characteristic trait of fast-track supervisors is their tendency to know how to get upper-management visibility for their employees. A fast-track supervisor will always have employees nominated for awards. Therefore you stand a much better chance of getting an award for your accomplishments when you work for a fast-track supervisor.

Once you are working for a fast-track supervisor you need to make yourself his or her right arm. You need to be alert and always ready to put in the extra effort demanded of you. You must give it your all and be loyal to and work as much as possible. Your primary objective is to make him look good so he will be promoted [27].

Of course, simply executing all his commands will not automatically result in promotions. You must demonstrate to him that you have the skills and knowledge to perform your job successfully. Your ability to get the job done and exhibit excellent work are basic requirements for successful career advancement.

Let's look at the negative aspects of settling for just any supervisor. Suppose you settle for a supervisor who has been in the same position for the last ten years and has shown no growth. This supervisor's career may have peaked and he or she has probably been passed over for promotion. He is as high in the company as he can go. Working for this supervisor could actually represent a roadblock to your career. If he is not moving up, you will more than likely also not be moving up either.

Another case in point is settling for an average performing supervisor who does not make waves. Supervisors who receive only average ratings tend to give only average ratings. Supervisors who want to only do their job and go home tend to stay away from upper management. If he is not getting upper-management visibility, then most likely you won't be either. These are only some of the examples as to why it is so important for you to select your supervisor and not settle for just anyone. A fast tracker quickly realizes the value in selecting his supervisor. If at all possible you want to choose your supervisor.

DEVELOP EXCELLENT COMMUNICATION AND PRESENTATION SKILLS

If you communicate in an average manner and your presentations are just average you can only expect to advance at an average rate. To get on the fast track you must have excellent communication skills and excellent presentation skills. The excellent communication skills are manifested in the written and oral reports you give to your supervisor and upper management.

Your written reports must clearly reflect your superior performance. All reports need to be well organized and neat in appearance. The results should be reported in a manner that allows the reader to clearly graph the significant conclusions. Graphical and mathematical analysis are a must. If possible the reports should at least be done on word processors and include color graphics.

Generating excellent written reports is not a natural talent. It is something that is acquired through practice and training. To learn impressive ways of writing you can take writing courses. I highly recommend you attend two types of writing courses. The first is a technical writing course and the second is an advertisement writing course. The technical writing course will show you how to organize technical material. The advertisement writing course will show you how to accentuate the positive and give it the additional flair to make it stand out from the rest.

A technical report can be extremely dry and hard to read. Even if your report represents a great scientific breakthrough, it still may get ignored due to poor writing. You need to jazz up your reports with a little advertising flair if you do not want your hard efforts ignored. The advertising course will show you how to do this. Learning how to write good technical reports with a bit of advertising flair is a must for the fast tracker.

Having great oral communication skills are even more important. You absolutely must know how to give a good oral report. Most upper-level managers do not have the time to read long reports. They want a summary of the significant points and they want it now. Oral reporting is also not a natural skill. You must develop this skill through training and practice. To develop this skill take a speech class or join Toastmasters. Toastmasters is an organization that meets on a periodic basis and its members practice giving speeches and oral reports. The organization provides step-by-step training and guidance on how to improve your presentations. Joining Toastmasters is an excellent way to develop your oral reporting talents and even meet upper level managers who often attend the group's meetings.

Practice is absolutely necessary to perfect your oral reports. One way to practice is by giving oral reports at home and use a video recorder to tape yourself and play back to critique yourself. You will be surprised at all the mistakes you make. By videotaping yourself you can quickly see what you have to work on. It's always better to make the mistakes at home than in front of your supervisor, co-workers, or at meetings.

Some people ask me if this is really necessary and do they need to practice. To this question I always respond with a large and emphatic *yes!* Athletes practice for hours; actors and actresses will rehearse their lines for hours; lawyers will practice their courtroom presentation; salesmen always practice their pitches; politicians practice their speeches; Clergy will practice their sermons. Why? To give the best performance they can. They want you to see them at their best. This is something natural that nearly all professions do. Not to practice is doing something that most other professions consider absolutely necessary. It would appear to be good common sense to practice any presentations you plan on giving. Why is it that engineers think they do not need to practice?

Finally, you must be able to make great presentations [14]. You must acquire the skills to speak in front of a group at meetings. For this you need to learn how to create presentation material, how to operate projectors, what makes a good chart, how to handle questions, and how to speak with authority. Great presentations are an absolute must for fast-track development. Remember, your supervisor will want the person with the best speaking and presentation skills to give the report to the vice president. Continue on if you're still interested in becoming a fast tracker.

GO FOR THE SPECIAL PROJECTS

Choose projects that will give you the most visibility. Projects that lock you in a lab with little or no visibility require special effort by the fast tracker to get that visibility. These projects may be really technically challenging but they can keep you away from the eyes of management. The fast tracker realizes that selecting your project is very important. If you have a choice of assignments, choose the supervisor's pet project. This is another fast-track move.

THE RIGHT TIMING FOR GETTING ON AND OFF PROJECTS

The fast tracker is said to be acutely aware that there are good times to get on a project and there are good times to get off a project. At the beginning of the project there is usually a lot of visibility by management to ensure the project gets started on the right foot. Upper management is normally involved and monitoring initial progress on the project. Getting on a project at the beginning provides the engineer with extra visibility. The beginning of a project is when most of the promotions are usually handed out. Therefore this is a good time to get on projects.

As the projects progress unforeseen problems may arise. These problems may result in schedule delays and large cost overruns. Most major problems will surface after about a year into the project. If you can see that the project is headed for disaster, the fast-track response will be to get off before the problems are discovered. This is an excellent time to leave a project. The fast tracker can honestly say the problems occurred after she or he left. This allows the fast tracker to be dissociated with any problem projects and always present a successful image.

When a project hits major problems, management often assembles teams to fix the problems. They sometimes refer to these as "Tiger Teams" or "Get Well Teams". The fast tracker will join a project that has severe problems only as a member of these Tiger or Get Well teams. Team members are usually assigned by upper management to solve the difficult problems. The fast tracker realizes the special teams have extra visibility with upper management due to the pressure to get the project back on track. Often the special teams must report daily progress to upper management. A fast-track engineer who has excellent reporting and presentation skills is now in a great position to move ahead quickly. This is the only good way and good time to get on a problem project.

As the problems are solved and the project comes to an end, the visibility of the project dramatically decreases with little opportunity for career advancement. However, engineers often continue to stay on the project with little or no benefit to their career. The engineer may stay too long because of the large amount of clean-up work needed to close out the project. Or the engineer may stay on the project longer than she or he should because the work was very interesting. Not wanting to leave the project is only natural, since you have spent so much time and effort getting the results. These are some of the reasons why the engineers stay on a project longer than is good for their career. The fast tracker is aware when a project has entered the final stage and has little or no benefit for his career. Fast trackers quickly move on to the next career-advancing project.

DRESS AND ACT THE PART

Study the dressing habits and mannerisms of your management. Determine the dress code and dress accordingly. Watch for mannerisms displayed by your supervisor. The fact of the matter is that supervisors tend to promote people who look like them and act like them. If it worked for them, then why not copy the successful formula?

SEEK OUT OPPORTUNITIES TO EXCEL

Opportunities to excel present themselves every day. Most people refer to these opportunities as big, unsolvable problems. Consequently they avoid them as much as they can. By avoiding them they lose the opportunity to shine. The fast tracker looks for these opportunities to excel and selectively chooses to work on them. Naturally, choosing the opportunities to excel that your supervisor considers the most important is key to rapid advancement. The fast tracker knows that the more desperate the boss is about the problem, the better the opportunity.

What if the opportunities to excel are on unsolvable problems? Don't worry, this is even better. The supervisor does not expect the employee to solve every problem but just his willingness to try is what will separate him or her from the rest.

MAKE YOUR OWN LUCK

The fast tracker realizes that most projects are not going to be highly successful and result in a promotion. As a matter of fact, the fast tracker realizes just the opposite, that most projects will experience severe problems and setbacks. She accepts this and plans for it. A fast tracker will always have backup or contingency plans. If you have backup plans that you can fall back on and keep the project moving ahead, it will appear that your project is successful.

> Making good backup plans is like planning for luck.

Stick to your plans and put in the effort to see things through to the end. Don't give up at the first sign of trouble. There is an old saying by Thomas Edison that best captures the definition of success. "Success is 99% perspiration and 1% inspiration." As a matter of fact, you can actually look on trouble as a good sign. If there are troubles they will need you all the more. This is called job security. With trouble comes extra visibility. With extra visibility and great performance comes promotion. Therefore trouble means job security and can even be the start of your next promotion. This should help you see the positive side of trouble and, hopefully, help you fear it less when it rears its ugly head.

These are only some of the tips that are available to you on fast tracking for engineers. Other sources of fast-track tips can be found in the trade journals, business magazines, and even professional society meeting proceedings [26,27]. You have to be alert and search for them. The best thing you can do is collect as many articles on the subject as you can. Place them in a binder and at least twice a year review the articles.

One final word about fast tracking. Sometimes engineers move up the corporate ladder at surprising speeds. Once in a while an engineer may move up two levels in a single promotion. This can be a great career advancement, but it can also be a fatal career move. With each level of promotion comes new responsibilities. At each level you need to develop the skills to perform at that level and make sure you can handle the responsibilities. If you move too fast and skip levels, you will not have the time to develop the necessary skills it takes to survive at upper levels. By skipping levels you more than likely will end up only demonstrating how incapable you are rather than how capable you are. To illustrate the point look at the following example.

A new student has just learned to swim. Since he has demonstrated that he can successfully swim the entire length of the pool, he feels he is now ready to swim the English channel. Even though he successfully swam in the pool, he is not yet ready for the English channel. Jumping multiple levels of engineering is similar to this example. You may feel confident of your ability to handle assignments two levels above you but you still must be careful that you don't accept something beyond your capabilities.

Experience has shown that there are reasons why people only get promoted one level at a time. If you start fast tracking and start to move rapidly make sure you can handle it. Give yourself time to learn the job skills required at each level.

HOW TO GET A 10^{27} IMPROVEMENT IN YOUR CAREER

People often ask what they can do to quickly accelerate their careers. What actions should they be constantly striving for that will most benefit their careers? To answer this question I have identified the value of various actions you can take to significantly increase your chances for career advancement. The following is a simplified priority action list for career advancement. By following this prioritized list of actions whenever you can, you should, hopefully, realize a 10^{27} improvement in your career.

Priority List for Career Advancement
($>10^{27}$)

One picture is worth 1000 words
One model is worth 1000 pictures
One successful project demonstration is worth 1000 models

One typed memo is worth 1000 hand written notes
One phone call is worth 1000 typed memos
One face-to-face meeting is worth 1000 phone calls

One phone invitation is worth 1000 typed meeting invitations
One face-to-face invitation is worth 1000 phone invitations
Free food at a meeting is worth 1000 face-to-face invitations

Bottom line: You realize a 10^{27} improvement in career advancement by having face-to-face meeting with upper management and provide free food while demonstrating your successful project.

CHAPTER 15

SUMMARY

Remember that it is not impossible to make changes, it's only difficult. The first step is getting control of your career. Controlling your career takes a great deal of hard work and managing circumstances the best you can for your benefit. You must *plan it!* Nobody is looking out for you and it's all up to you. Let us summarize the important messages you should have received from this book.

The ideas in this book only work if they are combined with excellent performance on the job, continuous improvement, commitment to do your best always and honestly. Your boss is not your enemy. Make your supervisor your ally. Make him look good and you yourself will look good in the eyes of upper management. This is the fastest way to get promoted. Your manager has been there—don't underestimate him or her. If you don't know, *ask!*

The first step to getting control of your career is determining the ladder structure of your company. What are the engineering, technical, and business ladders? Who controls your salary and who controls your work assignments? The product-oriented organizations offer advantages and disadvantages. The functional-matrix organizations also offer advantages and disadvantages. What hidden barriers exist in these ladder structures? Are advanced degrees a barrier; is the organization such that your educational background is limiting your career?

Next, understand the engineering process of your company. The engineering design, build, and test of a product is a complex process that involves a wide variety of people and departments. You must know how products are built and tested in the company. Determine if you are working for a mainline group. Identify the key departments and products of your company. When dealing with other departments create win–win situations. The engineering process continually changes due to improvements, department changes, and personnel changes. You must be constantly monitoring the changes and adjusting your actions appropriately.

If higher education is a must for career development, then use the benefits offered by your company's educational system. Find out the various types of educational courses offered by your company. There are in-house courses, week-long seminars, and more formal training available through university extension schools. Each of these training methods has advantages and disadvantages. Remember, continuing education is usually reimbursed by the company. Continuing education does not automatically result in a promotion; you must also maintain excellent performance.

If you feel trapped in a job position there is always the option of transferring out of the department. Your company should have a Job Opening System (JOS) that lists all the openings in the company. These lists are usually available through the personnel department. Using the JOS is a good means for career advancement, but you should be careful not to burn bridges when you leave. Leaving the company may be attractive and offer more money, but you may lose such benefits as vacation and seniority. You also might incur unexpected costs to change companies, making a smaller raise from your present company equivalent to a larger raise from a different company.

There are both formal and informal criteria by which you are judged. The company has predetermined formal criteria that might include promotion review boards and totems. Every job level has a formal description of the performance criteria necessary for that level. Your job appraisal forms describe your performance against the formal criteria. You must understand all the information that is summarized on the form. Your supervisor and human resources department will help you understand the forms and the formal criteria. Make sure you and your supervisor agree on the definition of your job. You must know the rules of the game if you expect to play well, score points, and get promoted.

The informal criteria by which you are judged are not always obvious. You need to be aware of the informal criteria and constantly checking for them. The informal criteria have a lot to do with your supervisor's styles and the accepted norms of your department. Informal criteria may include such things as neatness or appearance, style of presenting information (big picture versus little picture), methods of responding to problems, and office appearance. The informal criteria are not always obvious but they are there just the same. With a little research and acute observation you can recognize informal criteria and use them to your advantage.

Developing good mentors is a must. Mentors should be older, upper-level people who are willing to coach and guide you when you will encounter problems on the job. Mentors are very valuable resources who can identify solutions to difficult technical problems, provide career guidance, shelter you in times of trouble, or act as friends to bounce ideas off of.

Maintaining a company calendar is important. The company calendar allows you to see behind the scenes and identify when important activities that affect your career are occurring. It provides a forewarning and allows you to be well prepared in advance. Developing a company calendar is no easy task and it may take several years before you record all the significant company events. Once you do this, you will soon realize all the opportunities that are available to you.

Engineers are highly trained people with usually above-average intelligence. This does not necessarily guarantee a successful career. Studies have shown that the most common causes for engineers failing are: poor communications, poor relationships with the supervisors, inflexibility, poor work habits, and, finally, technical incompetence or obsolescence.

Simply having great ideas is not enough for the engineer. You must have the skills to sell your ideas to co-workers and your supervisor. The first step to selling your ideas is to get organized and write them down. Research the ideas and put together a well thought-out plan showing all the benefits and problems. Discuss it with your co-workers and get their inputs and backing. Finally give it your best shot and pitch it to management. Remember, if they do not accept it, there always is another idea.

You will encounter many fallacies in your engineering career. Blindly accepting these fallacies will greatly limit your career. Two fallacies not to be taken at face value are: successful projects result in promotion, and projects that fail will never result in promotions. In either case, you can still have career growth. To do so, you must enter a high-energy state and put in extra effort. For an unsuccessful project you need to get organized, develop a recovery plan, and brief management on reasons for the failures.

For a successful project you need to get organized, brief management on the good results, and identify reasons for the success. Excellent technical presentations showing theory, modeling, and test results are a must for both.

Finally, career development is a never-ending task. If you are not growing and expanding your capabilities you are decaying and slowing becoming obsolete. Career development is something that you must work at every day. A promotion in most cases is a year away from the time you start planning it. So if you want a promotion next year start planning for it now so when the time comes you will have fulfilled all the criteria. Remember, your raise only becomes effective after you do!

What does it take for successful career development?

It's *what you know, who you know, excellent performance,* and *how well you present yourself.*

A final note in closing. I would like to receive feedback from you on how the hints presented in this book are working. I would like to hear from you about what has worked for you and any other ideas you may have discovered for engineering career development. Please limit your responses to one page and send your comments to:

> John Wiley & Sons
> Interscience Division
> 605 Third Avenue
> New York, NY 10158-0012
> Attn: Career Advancement and Survival for Engineers

REFERENCES

1. B. Kleiner, "Managing Your Career," *Supervisory Management Journal,* March 1980, 15–21.
2. M. Badawy, *Developing Managerial Skills in Engineers and Scientists,* New York: Van Nostrand Reinhold, 1982.
3. J. Stoner, *Management,* Englewood Cliffs: Prentice-Hall, 1982.
4. G. Odiorne, *How Managers Make Things Happen,* Englewood Cliffs: Prentice-Hall, 1974.
5. R. Fisher, and W. Ury, *Getting to Yes,* New York: Penguin Books, 1983.
6. M. Badawy, "Finding and Using a Mentor," *Machine Design Magazine,* Aug. 9, 1984, 65.
7. J. McAlister, "Why Engineers Fail," *Machine Design Magazine,* Feb. 23, 1984, 47.
8. E. Raudsepp, "How Engineers Get Ahead and Avoid Obsolescence," *Machine Design Magazine,* Jan. 12, 1986, 107.
9. J. Nirenberg, *How to Sell Your Ideas,* New York: McGraw-Hill, 1984.
10. N. Hill, *Think and Grow Rich,* North Hollywood: Wilshire, 1973.
11. L. Giblin, *How You Can Have Confidence and Power,* North Hollywood: Wilshire, 1956.
12. D. Fuller, "How to Write Reports that Won't Be Ignored," *Machine Design Magazine,* Jan 11, 1979, 76–79.
13. E. Raudsepp, "The Politics of Selling Ideas," *Machine Design Magazine,* Nov. 7, 1985, 97.
14. R. Kliem, "Making Presentations that Command Attention," *Machine Design Magazine,* April 9, 1987, 143.

15. R. Kirkham, "Getting the Boss to Accept Your Ideas," *Machine Design Magazine,* October 22, 1987, 153.
16. A. McGinnis, *Bringing Out the Best in People,* Minneapolis: Augsburg, 1985.
17. L. Lucanio, "How to Get Job Security," *Money,* Feb. 1992, 57.
18. J. Shandle, "Engineers Have to Adapt," *Electronics,* Feb. 1992, 32.
19. N. Better, "Beating the Recession Odds," *Glamour Magazine,* April 1991, 310.
20. R. Schuller, *"Success is Never Ending, Failure is Never Final,"* New York: Bantam Books, 1980.
21. D. Lacey, "Employee Rights in a Rotten Economy," *US News & World Report,* Feb. 24, 1992, 70.
22. B. Nussbaum, "A Career Survival Kit," *Business Week,* Oct. 7, 1991, 98.
23. E. Raudsepp, "Plotting a Course For Career Success," *Machine Design Magazine,* March 22, 1984, 51.
24. E. Raudsepp, "Do-It-Yourself Career Planning," *Machine Design Magazine,* April 11, 1985, 99.
25. J. Cowle, *"How to Survive Getting Fired and Win,"* Chicago: Follett, 1979.
26. P. Sweeney, G. Dwyer, "Getting on the Fast Track," *Machine Design Magazine,* Nov. 10, 1988, 49.
27. R. McGarvey, "Working with the Boss," *USAir Magazine,* Sept., 1992.
28. R. Ringer, *Looking Out For #1,* New York: Fawcett Crest, 1977.

INDEX

193